60th Annual Symposium of the Society of Experimental Test Pilots (SETP 2016)

Anaheim, California, USA
21 – 24 September 2016

ISBN: 978-1-5108-3751-5

Printed from e-media with permission by:

Curran Associates, Inc.
57 Morehouse Lane
Red Hook, NY 12571

Some format issues inherent in the e-media version may also appear in this print version.

Printed by Curran Associates, Inc. (2017)

For permission requests, please contact Society of Experimental Test Pilots
at the address below.

Society of Experimental Test Pilots
P.O. Box 986
Lancaster, CA 93584-0986
USA

Phone: 661.942.9574
Fax: 661.940.0398

Setp@setp.org

Additional copies of this publication are available from:

Curran Associates, Inc.
57 Morehouse Lane
Red Hook, NY 12571 USA
Phone: 845-758-0400
Fax: 845-758-2633
Email: curran@proceedings.com
Web: www.proceedings.com

TABLE OF CONTENTS

Aerial Refueling Probe Oscillation Testing of the HH-60G Pave Hawk

Major Derek Spear
Ms. Julie Clark
Combat Search and Rescue Combined Test Force (CSAR CTF)
Nellis AFB, NV 89191

Abstract

The United States Air Force HH-60G Pave Hawk fleet has had several reported incidents of extreme oscillation of the aerial refueling probe when extended in flight. This has resulted in structural damage to the probe and surrounding airframe, and in one case the refueling probe departing the aircraft in flight. The HH-60G System Program Office and the original equipment manufacturer had analyzed several of these probes post incident, but the cause of the oscillations remained unclear. Following the post-flight analysis efforts, a more direct investigation of the phenomenon through flight and ground testing commenced.

The aim of this paper is to describe the test process and methods used during flight testing to characterize the HH-60G refueling probe oscillation phenomenon, to include initiating the oscillation through the Automatic Flight Control System, as well as discuss the structural and aerodynamic correlations presented in the test results and lessons learned from a joint government-contractor test program. This is done using a lessons learned focus on aspects that are relevant across all test programs.

Introduction

The US Air Force HH-60G fleet has had several reported incidents of extreme oscillation of the aerial refueling probe when extended in flight, resulting in structural damage to the probe and surrounding structure and, in one case, the probe departing the aircraft. As a safety of flight issue, the HH-60G System Program Office (SPO) advised any flight crew encountering this condition to annotate the incident in the aircraft forms for tracking information. The HH-60G SPO and the original equipment manufacturer (Sikorsky Aircraft Corporation) analyzed several probes post-incident, but the cause of the oscillations remained unclear. With no further useful data to be gained from post-incident analysis, the HH-60G SPO requested a more direct investigation of the phenomenon.

In March of 2012, Sikorsky and the Combat Search and Rescue Combined Test Force (CSAR CTF) executed initial testing for the probe oscillation issue resulting in 42 ground test hours and 13.8 hours of flight testing. Data from three aircraft were collected from steady state airspeed sweeps, Stability Augmentation System (SAS) step inputs, pilot pulse inputs, and SAS sine wave frequency sweeps. Additionally, some testing was completed with reduced rotor RPM (Nr). Although data were collected from this test, full data analysis was not completed until December 2014 due to resource constraints.

The results of the 2012 testing showed significant radial free play on the refueling probes that oscillated. Additionally, the results showed that a weakened cockpit tub and probe attachment structure may have made probes more susceptible to oscillations through a flight control interaction.

Between the completion of joint Sikorsky and CSAR CTF flight testing in 2012 and the initiation of ground testing for the latest test effort in 2015, Mercer Engineering Research Center (MERC) was tasked to analyze refueling probe oscillation problems on the HH-60G and to develop a solution that could be implemented on the Air Force fleet. MERC identified a flexing tub structure under the pilot's floor leading to the discovery of movement of the cyclic pitch control bell crank mount. This movement was thought to translate to swashplate and rotor movement, potentially providing a feedback loop to allow or sustain probe oscillations. As a result, a subfloor repair was developed to re-strengthen this floor structure and prevent the feedback loop.

This repair was not considered a solution to the probe oscillation issue, although it was considered a mitigating factor.

Testing for the 2015 effort began with ground testing at Gabreski Air National Guard Base, NY on 18 February 2015. Resonance Assessment Profile, radial play, probe shaker, and cyclic shaker tests were performed on a standard fleet refueling probe with no history of oscillations ("non-oscillator") and refueling probes that had previous oscillations in flight ("known-oscillators"). Six of the known- oscillators were rebuilt according to the Air Force aerial refueling probe Technical Order rebuild process ("rebuilt probes"). These rebuilt probes, along with the non-oscillator and the remaining known-oscillators, were retested on the ground. Additionally, one brand new probe with zero flight hours ("new probe") was ground tested in the same manner as the other probes. A grand total of 110.8 ground test hours were executed for this ground test phase. While previous testing pointed to radial play as the primary contributor of probe oscillations, the February 2015 ground testing did not directly support that theory, producing a more complex picture of the issue, as well as calling into question the Air Force rebuild process.

The Fall 2015 flight test evaluation was designed to evaluate both aircraft and probe factors as potential causes. Four probes were installed and tested in turn on the test aircraft. The test team performed a short series of ground test events before attempting to collect vibration data in flight. Another round of the ground test events were performed to identify any structural changes. Following a large limit-cycle oscillation, flight testing was put on pause to determine the next step forward.

In Spring 2016, a follow-up round of flight testing using the same test points was performed on a probe that had been rebuilt with characteristics thought to reduce the likelihood of oscillation, while using a modified floor plate to increase the stiffness of the floor.

Test Aircraft

The HH-60G Pave Hawk helicopter (Figure 1) is a modified version of the UH-60 Blackhawk manufactured by Sikorsky Aircraft Corporation in Stratford, Connecticut. The primary mission of the helicopter is Combat Search and Rescue and the primary aircrew of four includes the pilot, copilot, and two enlisted aviators called Special Mission Aviators (SMAs). Up to three pararescuemen are integrated into the crew for rescue operations.

Figure 1. HH-60G Aerial Refueling (U.S. Air Force photo by Staff Sergeant Caleb Pierce)

The basic Blackhawk has a semi-monocoque fuselage and a single four-bladed, fully articulated main rotor. The HH-60G has been modified from basic Blackhawk to include a refueling probe for in-flight refueling operations.

The HH-60G is also equipped with an Automatic Flight Control System (AFCS) comprised of an analog SAS (SAS 1), a digital SAS (SAS 2), Flight Path Stabilization (FPS), trim system, Improved Altitude Hold and Hover Stabilization (IAHHS), and stabilator system. The SASs provide short-term rate damping in roll, pitch, and yaw axes and can be engaged or disengaged independently via the AFCS switch panel (SAS1 and SAS2 switches). Each SAS has a 5% control authority, providing a total SAS control authority of 10%. The trim system, which can also be engaged and disengaged by the AFCS switch panel (TRIM switch), maintains position of the cyclic and tail rotor controls. The FPS provides long-term rate damping in pitch, and can be disengaged via the AFCS switch panel (FPS switch).

Test System

The aerial refueling system on the HH-60G uses an extendable refueling probe, which is hydraulically driven with fuel pressure from the fuel transfer pumps.

The probe is located on the right side of the aircraft and is comprised of a fixed outer tube assembly, an extendable tube assembly with nozzle, forward and aft bearings, and various fuel lines and miscellaneous parts (see Figure 2 and Figure 3). The probe is attached to the aircraft in two locations: the aft probe attachment point is to a plate just aft and below the pilot door and the forward probe attachment point uses a banjo fitting bolted to a plate extending from the floor beam where the cyclic controls are mounted (see Figure 4).

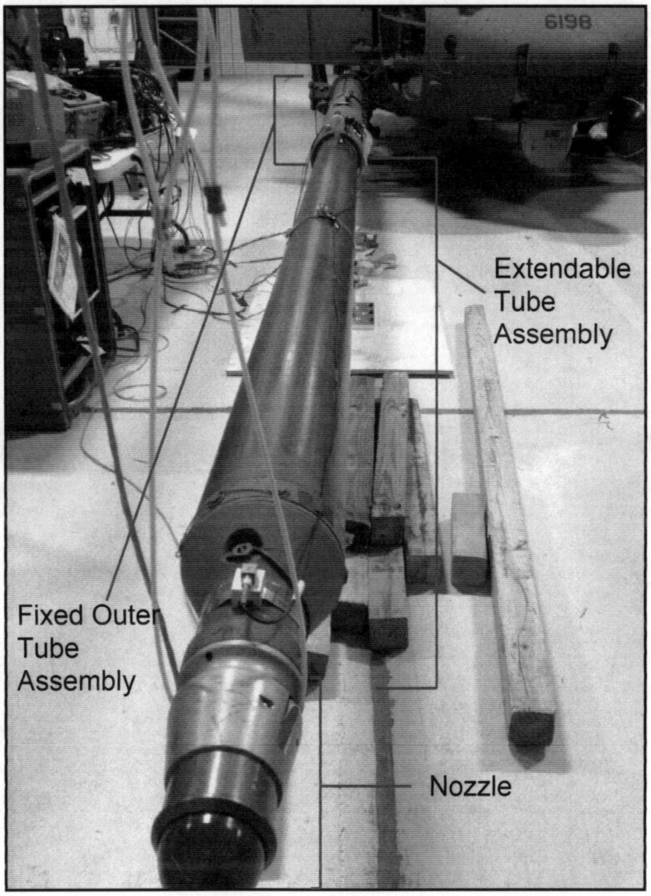

Figure 2. Fully Extended Refueling Probe

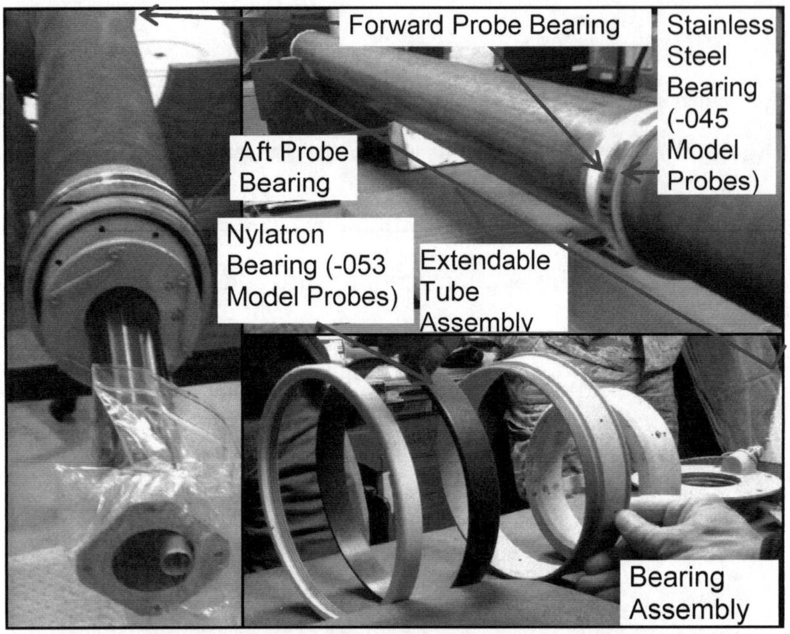

Figure 3. Disassembled Probe Structure

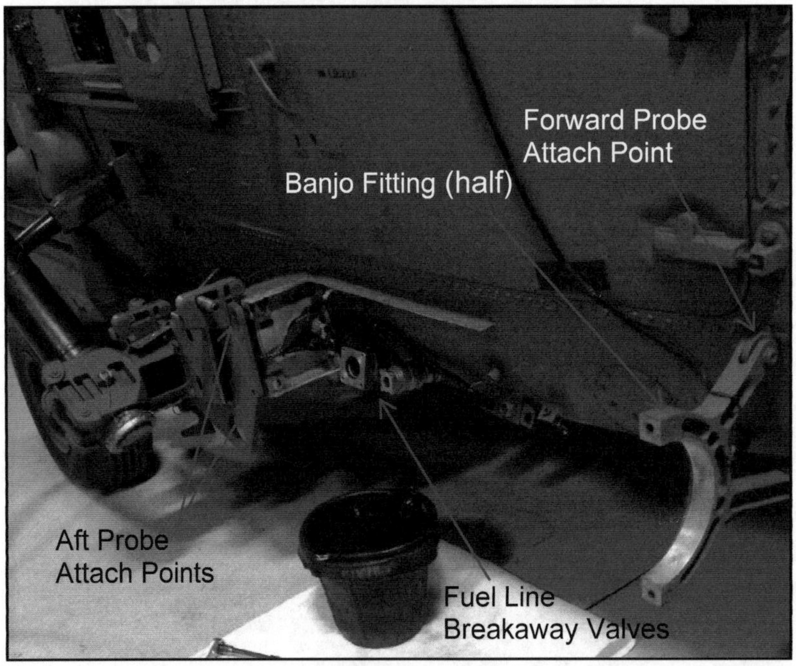

Figure 4. Probe Attachment Locations

One of the major technical difficulties with testing the refueling probe issue stemmed from the wide variety of internal probe configurations. Design changes throughout the life of the HH-60G resulted in a variety of fielded configurations consisting of these part options: Nylatron (plastic) bearings or stainless steel bearings; multiple bearing sizes; split versus un-split bearings; lined, unlined, and lining removed tubes; original Sikorsky manufactured probes versus later produced probes; and rebuilt versus original probes.

Additional configuration issues resulting from the state of individual aircraft were also a concern. Each HH-60G airframe has a wide variety of flight hours, previous structural repairs, major depot level repairs, and different aircraft uses.

Method of Test (MOT)

Because probe oscillations were a problem to the HH-60G fleet for over 10 years, a large pool of data were eventually collected on the conditions surrounding oscillations. Aircraft flight hours, basic flight conditions, probe state, aircraft location, and pilot on controls during the event, and aircrew comments were collected. Aircraft of a wide range of flight hours had oscillations. Probes both extended and retracted had oscillations, although extended probes were the vast majority. Several of the historical incidents indicated that oscillations were more likely to occur if the right-seat pilot was on the controls vice the left-seat. Oscillations seemed to happen more frequently in the refueling pattern with the tanker or in pre-contact position. The tanker drogue was also noted to induce oscillations when it contacted the probe or on separation in some cases. Other pilots indicated that certain SAS and TRIM configurations of the AFCS impacted the likelihood of oscillations. The documentation in many cases were incomplete and many contained speculation on the cause.

Overall, the historical data failed to show any significant trends with one exception: Kirtland AFB, the training base for all HH-60G aircrew, had the highest percentage of oscillations to flight hours by a wide margin. Because Kirtland AFB is a very unique base for the HH-60G, its aircraft are also unique in use and age. Kirtland aircraft have some of the highest time airframes in the fleet. These aircraft are often flown by inexperienced pilots, but are also flown by some of the most experienced pilots in the fleet. The aircraft often have harder landings and are used for aerial refueling more than at any other location. Without detailed information about each event at Kirtland, it was impossible to determine which factors were relevant to probe oscillations.

Because of the very large variety of options to test with only three clues to the cause (loose bearings from initial flight testing, indications of low root end stiffness from ground testing, and historical data indicating a high percentage of oscillations at Kirtland AFB), many potential factors were grouped for testing. The variety of probe internal configurations would result in a higher or lower radial free play. Probe manufacturing conditions and conditions affecting probe wear and tear would result in a lower or higher root end stiffness value. Two major manufacturing phases broke out the design of the probe into two specific models, the -045 probe and the -053 probes. AFCS configurations, airspeed, pilot flying, and pilot method of control of the cyclic were also selected as factors for testing based on information in the historical data. With a wide variety of airframe characteristics that could contribute to oscillations, one aircraft with no history of probe oscillations was chosen for testing to focus on probe specific causes of oscillations.

As a result of many oscillation events in the fleet, aerial refueling probes with oscillations were quarantined for future testing or rebuilding through the Air Force rebuild process. Ground testing was performed to accurately measure a wide variety of characteristics of each available quarantined probe (the same testing that identified the trend in the root end stiffness). Of these ground tested probes, several were selected that had extremes in factors relevant for the test. One standard, fleet comparable probe with no history of oscillations and one brand-new, never flown probe were also chosen for testing for comparison.

In addition to the factors chosen for testing the cause of oscillations, other factors had to be included for data collection (frequency sweeps versus frequency dwells), data trends and safety buildup (input magnitude in steps of 0.5% and probe retracted then extended), and initial aircraft health. Even with the factors to test greatly reduced, the test matrix with probe serial number, AFCS configuration, Pilot flying and Control Method, Altitude, Airspeed, test point type, probe condition (original versus rebuilt) contained 1,987 flight test points. Those points did not even account for ground test points both before and after each flight to identify structural probe/aircraft changes.

There were several constraints that ended up limiting the number of test points to complete during testing. The maximum input amplitude for each of the test points was capped by the maximum load (via accelerometers) on the probe for each successive amplitude determined to be safe for testing. AFCS configurations that showed little response while testing the first few probes were eliminated later in testing for other probes. Benign responses for pilot control and flight technique for a given probe were used to eliminate test points at different airspeeds and input amplitudes. In spite of the large oscillation that ended flight testing early, nearly two-thirds of testing was either completed or eliminated.

Safety

Safety was the primary concern when probe oscillation testing was being planned. A very specific buildup method was developed in an attempt to slowly approach any instabilities and attempt to identify trends in probe behavior. Initially, the aircraft was flown in straight and level flight with a full suite of accelerometers to characterize the baseline airframe behavior in normal flight. This set of test points at 6 different airspeeds was completed and reviewed before any attempts were made to induce a frequency response from refueling probes. After the aircraft health was verified, a set of test points of frequency sweeps and dwells was completed with a non-oscillator installed and fully retracted, the configuration least likely to oscillate while still sending input signals to the AFCS. Testing progressed from using a non-oscillator to known-oscillators, from hover to low forward speed to high forward speed, from less likely Pilot-in-the-loop oscillation (PIO) AFCS configurations to more likely PIO AFCS configurations, right seat to left seat, and low amplitude inputs to high amplitude inputs. Intermediate buildup test points were added when trends were not clear or when stepping up to the next higher amplitude would likely exceed test specific limits. Throughout testing, probe and aircraft behavior were monitored for trends and buildup order was adjusted to always proceed from more benign to less benign conditions.

With the nature of structural testing and the tendency towards the "structural cliff" where behavior rapidly changes from stable to unstable, the test team used both test "abort" values and safety knock-it-off ("KIO") values.

A safety KIO value of 19.2Gs of probe tip acceleration was set based on 50% of a probe attachment structure peak Do Not Exceed (DNE) value, meaning that the probe could oscillate at twice the acceleration or load before experiencing damage. This value was also compared to estimate probe tip deflection levels (deflection converted to load then acceleration based on structural beam bending) in fleet probe oscillations, which supported the notion that aircraft damage would not occur at this value. If a KIO were reached, all AFCS inputs would be stopped and then disconnected via the AFCS training simulator, the pilot would reduce airspeed and lightly load the probe by initiating a level turn, retract the probe when able, and return the AFCS to a normal configuration in that order. Testing would then cease, a Battle Damage Assessment (BDA) would occur by the chase aircraft, and then Return to Base (RTB) for a post-oscillation inspection.

A test abort limit was set at 7.3Gs to prevent the accidental exceedance of the KIO value due to continued oscillation growth post-input termination. This value was set based on the 30 minute vibration fatigue limit assuming a probe 3.5HZ natural frequency and the peak DNE. Since probe response decay behavior was needed for this test to help predict approaching instabilities, residual "ringing" was expected. Therefore, when the test abort limit was reached, the AFCS input was terminated and disconnected via the AFCS training simulator, but no other corrective actions were taken as no damage was expected at this lower limit.

As with most structural flight testing, a safety chase aircraft was used for ground, tower, and area comms, situational awareness, visual test aircraft monitoring by sight and with a videographer, altitude monitoring, and visual test aircraft inspection after specific test points to assess any aircraft damage (i.e. BDA).

Development of a chase training plan was identified as a serious challenge at the beginning of the test planning phase as the CSAR CTF is a combined Air Combat Command (ACC) and Air Force Material Command (AFMC) flying unit that typically operates under ACC flight rules. However, ACC does not define a safety chase aircraft in its regulations as ACC does not typically do Developmental Test (DT). Additionally, as the test was expected to be a MEDIUM or HIGH risk test, AFMC regulations dictate that mishap responsibility must be transferred to AFMC and AFMC flight rules apply. Given that the test aircraft had been specifically borrowed from a National Guard unit for this test effort and grounded by the SPO under AFMC to maintain the exact same structural integrity between ground test and flight test phases (several months apart), the HH-60G SPO was unwilling to allow ACC to fly the aircraft for anything but probe testing. Since the CSAR CTF only owns one aircraft, operates under ACC flight rules, and the SPO limiting the use of the borrowed test aircraft for when the test actually began (meaning all test safety paperwork also had to be obeyed), chase training was relegated to the two weeks prior to actual probe testing (officially operating as a MEDIUM risk AFMC test).

While AFMC defines the purpose of a safety chase aircraft and defines qualifications for safety chase for most fixed-wing platforms, no qualification information was provided on the HH-60G. The chase qualifications of one pilot in the H-1 from his prior test assignment were used to define the first HH-60G qualified chase pilot. The formation qualifications and training in the HH-60 under ACC were used as the basis for training and qualifying pilots for safety chase under AFMC. The intent of ACC regulations also conflicted with the intent of AFMC regulations in other ways. ACC defines the minimum primary aircrew for the HH-60G as 3 persons with a standard primary aircrew of 4 based on mission complexity. A complex mission like probe testing required a primary aircrew of 4 for minimum manning. The AFMC stance on aircrew is to minimize the loss of life in the event of a mishap under test. The test team spent a significant amount of time justifying 4 primary flight crew, 1 Air Force qualified Test Conductor (TC), and the 3 Sikorsky engineers flying as Mission Essential Support Personnel (MESP) for the technical aspects of the test.

As the test required a total of 4 pilots for every flight and Sikorsky required that one of their pilots be on the test aircraft to meet their safety requirements, a Sikorsky test pilot flew every test flight. Three Sikorsky MESP were also required by Sikorsky to operate the oscillator box providing the input, monitor accelerometer outputs for abort and KIO limits, and verify data quality. While test limits and data quality verification is typically performed by a qualified Air Force TC, Sikorsky refused to perform the test without their personnel controlling those aspects of the test. With 4 contractors onboard with no Air Force training, both Defense Contract Management Agency (DCMA) regulations for contractors in flight with approval from a Government Flight Representative (GFR) had to be met and general familiarity training had to be accomplished.

Training

General familiarity training was accomplished for all onboard aircrew through academics and multiple local familiarization flights in conjunction with chase qualification training. These flights for the contractors included familiarity with Air Force personal equipment storage and restraint system usage, Air Force specific cockpit layout, local area landmarks and hospital locations, flight in the reserved probe test track (an aerial refueling track was reserved through the Notices to Airmen [NOTAM] System), and emergency landing locations. Although the

Sikorsky test pilot was an extremely qualified pilot in the base H-60, he was not specifically qualified in the HH-60G aircraft, which is not a DCMA requirement. This pilot spent considerable time during the two weeks leading up to probe testing becoming familiar with the flight characteristics and nuances of this unique model of aircraft.

Communications (Comm) and Crew Resource Management (CRM)

Due to the mixed onboard aircrew, special emphasis was placed on developing a strict comm plan to combine the unique aspects of HH-60G CRM, test specific comms through the designated Air Force TC, and Sikorsky's less regulated test comm style. HH-60 CRM involves a looser comm style than typical fighter aircraft brevity comm as the primary flight crew is larger and the SMA role in keeping the aircraft safe in a variety of flight conditions including takeoff, landing, approach, turns, and hover is critical and mandatory. While the ACC aircrew was familiar with operating with an Air Force TC and the qualification of this aircrew member is the same under ACC (as an Operational Test [OT] TC) as it is under AFMC (as a DT TC), OT TCs are typically providers of information to the pilots who run the test as opposed to DT TCs that direct the pilots through their actions. This difference stems from the different mission of OT versus DT: OT tests equipment and tactics using operationally representative maneuvers and scenarios where the primary aircrew are experts, while DT tests equipment and systems to characterize behavior and identify proper operation where the engineers are the experts. Since Sikorsky typically performs acceptance tests and general aircraft troubleshooting, their comm style was more discussion based and focused more on technical information than the delivery of safety-related information. The delineation between test safety responsibility, aircraft and aircrew safety responsibility, and technical data collection requirements was very blurred for this test.

The final comm and CRM plan involved several moving parts. The chase aircraft maintained responsibility for all external, non-test related comms and was lead aircraft going to and from the test airspace, as well as during emergencies. After being led to the test airspace the test aircraft took lead and the chase aircraft took chase positioning on the right side in a slight trail position for best observation of the refueling probe. The chase aircraft monitored the local airspace calls including test external radio calls during actual test point execution, radioing the test aircraft with wind updates, traffic calls, turn coordination at the end of the airspace, test time and fuel calls, and acknowledgement of the beginning and ending of test points. The test aircraft focused on test specific flight parameters and test aircraft safety.

To help aid everyone in clearly understanding everything that was being communicated, a set of definitions for various calls and terms were defined specifically for this test within the front matter of the test cards. The following terms were used:

Abort (repeated 3 times): Test limit call. Test point execution should immediately be ended, however aircraft recovery is not required unless the situation transitions to a safety of flight issue.

Knock-It-Off (repeated 3 times): Safety of flight call. Test point execution should immediately be ended and the planned recovery procedures should be completed immediately. This call resulted in an RTB if the call was related to the probe testing itself.

Terminate: Test point execution should be ended as continued execution will not provide useful data.

Sterile Cockpit: End all non-critical comms immediately.

Beginning: Inputs are being injected into the AFCS.

AFCS ON: The AFCS training simulator is being switched to allow injects in the AFCS. Any residual or accidental inputs from the sine wave input box may now be transmitted to the AFCS.

AFCS OFF: The AFCS training simulator is being switched to prevent any injects in the AFCS. Any residual or accidental inputs from the sine wave input box will not be transmitted to the AFCS.

Ready: The aircrew member has properly configured all equipment and is immediately ready to monitor and record data.

On conditions: The Pilot Flying (PF) has stabilized on the proper setup conditions and is ready for AFCS signal injects.

Data off: Technical data call. Data recording is being turned off.

Test Point Complete: All required data were required for the test point and the test will progress to the next test point.

Sheldon: The callsign designated for the TC as the TC changed from flight to flight.

Box: The callsign for the MESP controlling the sine wave input box.

Simulator: The callsign for the SMA controlling the AFCS training simulator.

Onboard the test aircraft, the PF was responsible for listening to the TC ("Sheldon") for initial setup conditions and then maneuvering the aircraft onto test parameters. The Pilot Not Flying (PNF) was then responsible for listening to "Sheldon" for the AFCS configuration and then setting the proper switches. After the aircraft was configured, the PF would make small adjustments to regain the proper test point setup that was sometimes disturbed by engaging/disengaging automatic control of certain aspects of the flight control system. Once ready to begin the test point execution, the PF simulcasted internally to the crew and externally to chase that he was "On Conditions". "Ready" calls internally from "Sheldon" and the MESP monitoring the abort and KIO parameters were then made. The SMA controlling the AFCS training simulator ("Simulator") would internally call "AFCS ON" to indicate that external inputs to the aircraft flight control computer were now possible. The MESP controlling the sine wave input ("Box") would then indicate the setting for the test point amplitude and then simultaneously speak and slowly dial up the input frequency from 2 Hz to 5 Hz during frequency sweeps or indicate the test frequency and say "Beginning" for frequency dwells. "Sheldon" monitored and made time hack calls during the frequency dwells. On completion of a test point, "Box" would call "data off", the data quality MESP would also call "Data Off" for a second set of equipment, "Simulator" would call "AFCS OFF" and turn off all possible input to the AFCS, followed by a "Terminate" call by "Sheldon", echoed internally and externally to chase by the PF. The calls by "Sheldon" were intended to be the only calls of interest to the pilots (except test abort calls or KIO calls by any aircrew member) to provide a point of focus in a comm saturated environment. After execution of a test point, the data were reviewed and discussed and a determination of test point completion was made. "Sheldon" relayed the determination to the PF through a "Test Point Complete" call.

Results and Test Termination

A total of 272 test points were executed over the course of 9 test flights with 3 aerial refueling probes. While the team expected to see some distinct trends in probe response with some distinct factors, the responses seen did not correlate from probe to probe and airspeed to airspeed. Higher responses were typically consistent with one AFCS configuration, but that configuration changed from probe to probe. Testing performed in a hover produced consistently less response than testing in forward flight, but the response was not consistent from probe to probe across the two forward airspeeds. The only relatively consistent trend across the entire test was increasing the amplitude input resulted in a higher amplitude response, although there were a few outlier test points as well. The team was able to somewhat predict if the next test point would exceed the abort criteria in forward flight based on hover testing results combined with the trend in response for a given configuration.

The physical response of each of the probes in variety of test conditions varied considerably. The probes tended to flop in certain conditions, ring in others, stay heavily damped in others, and have a high response but then quickly damp in yet other conditions. At some conditions, the probes would clunk loudly where the probe bearings met the outer cylinder while other times they were silent.

During the ninth flight of the test program on the fifth test point of the day, the "structural cliff" was reached and a large limit-cycle oscillation occurred. The following table outlines the completed test points of the day with their peak probe tip accelerations and the resulting acceleration for the limit-cycle oscillation.

Test Point	Time	Description	AFCS Config	Pilot Flying and Control Method	Peak Tip Acceleration (G)
IPFT480	18:42:29	1% SAS1 Sine Sweep 2-5Hz, 4500 ft MSL, 110 KIAS	SAS off, Trim off, FPS off	Either, Guard	3.68
IPFT477	18:44:10	1% SAS1 Sine Sweep 2-5Hz, 4500 ft MSL, 110 KIAS	SAS on, Trim on, FPS on	Either, Guard	7.07
IPFT473	18:46:21	1% SAS1 Sine Sweep 2-5Hz, 4500 ft MSL, 110 KIAS	SAS on, Trim on, FPS on	Copilot, Fixed	4.88
IPFT472	18:49:12	1% SAS1 Sine Sweep 2-5Hz, 4500 ft MSL, 110 KIAS	SAS off, Trim off, FPS off	Pilot, Fixed	3.74
IPFT469	18:50:51	1% SAS1 Sine Sweep 2-5Hz, 4500 ft MSL, 110 KIAS	SAS on, Trim on, FPS on	Pilot, Fixed	22.37

As can be seen from the progression of test points through various configurations, there was no indication that the "structural cliff" existed. Even though there was a higher response when using the AFCS all on configuration, there was no upwards trend when switching from the Either, Guard Control method to the Pilot, Fixed method.

The following sequence of events and associated figure are taken from the Quick Look Report generated within 2 days of the oscillation event that illustrates the sequence of events that occurred on that test point:

> **Card 33:** The abort and subsequent KIO on test point IPFT469 due to probe oscillation in excess of 19.2G was not expected, given the benign nature of the previous test points. As the sweep approached the resonant frequency of the probe, tip accelerations grew steadily to the abort criteria. Abort was called by the Sikorsky Engineer as tip accelerations clearly passed 7.3G and the call was echoed by the TC. Tip accelerations continued to grow rapidly over the next 5 seconds to approximately 21G. Aircraft vibration and aircraft flexure increased dramatically as probe amplitude built and then remained steady in a 22G limit cycle oscillation at a 3.5 Hz frequency. At the abort call, the AFCS training simulator, the device transmitting the sine input to the SAS, was immediately turned off. Both pilots waited a few seconds to ascertain if the probe was damping before initiating additional corrective actions. As the oscillation grew in amplitude, a deceleration of approximately 2 kts/sec was initiated. At approximately 70 KIAS, the oscillation reached a steady state and the pilot-not-flying initiated a probe retraction. Despite verifying a refueling probe retraction rate of 17 seconds on the ground, the retraction rate was extremely slow (nearly 3 minutes) during the oscillation. Aircraft deceleration at a rate of 2 knots per second continued to 50 KIAS. The oscillation remained at its peak for approximately 7 seconds before decaying over a period of 30 seconds. During the decay, the pilot loosened his grip to reduce his feedback to the flight controls. The pilot flying was just about to pass the controls had the oscillation not clearly dampened. The following chart shows the probe tip acceleration throughout the event with notes about times during the event.

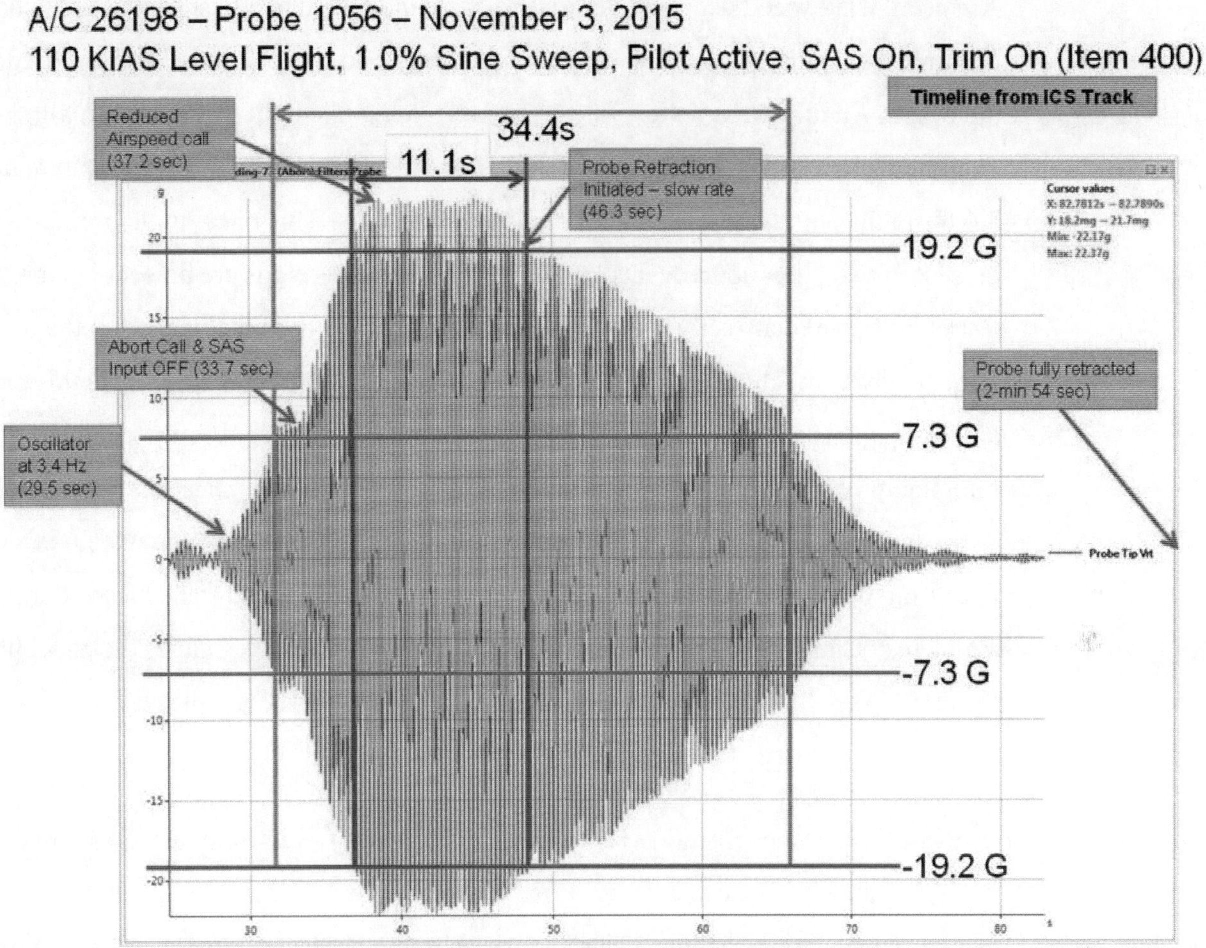

Figure 5. Refueling Probe Tip Accelerations

Further testing was terminated in accordance with the Divergent Refueling Probe Oscillation Test Hazard Analysis [i.e. safety planning. A BDA was completed by the chase aircraft and revealed no discernible damage. Due to a fuel odor in the cockpit, the two aircraft initiated a precautionary landing at a local airport. Upon landing, the aircrew noted a fuel leak of approximately 1 drip/sec coming from the root end of the probe. Upon applying the rotor break, the pilot noted a cracked upper window. After the engines were shutdown, maintenance performed a post-oscillation inspection (see MOT Annex 3) according to THA requirements under supervision from the SPO engineer that developed the procedures resulting in zero findings of structural damage.

A thorough review of available bus data revealed no discernible trigger to this event, nor any indicator that an oscillation would grow to the amplitude seen in this event.

Airspeed, wind speed data, and vertical velocity indicate that a gust of wind may have occurred during the sine sweep, although the timing does not indicate a direct trigger of the event. Additionally, review of the aircraft pitch angle indicates a very small amplitude pitch angle oscillation at 3.4 Hz from the point where the SAS input reaches 3.6 Hz to the point where the pilot begins to decelerate. This pitch angle oscillation is not present in two other abort events, although discontinuous signs are present in a previous test point flown incorrectly using the fixed control method; in that test point the frequency was much further from the natural frequency of the probe. While this pitch angle phenomenon may indicate pilot feedback or potentially aircraft excitement of the refueling probe, implementation of an abort call based on this indication is infeasible. Also, due to the data review, the flight test engineers noted a discrepancy between the pilot's airspeed indicator and the airspeed data from the airspeed transducer. Further maintenance inspections found that the transducer, which resides under the panel under the pilot's seat, was not operating within specifications and was replaced.

The data as analyzed thus far does not support a definitive conclusion as to the root cause of the probe oscillation. Hypotheses have been offered from several sources regarding the ultimate cause of what makes a probe an "oscillator" and what actually causes an oscillation in flight. The Sikorsky contractors have stated that they have sufficient data to perform an adequate analysis to determine the root cause.

The sudden growth of the oscillation from approximately 7G (when abort was called) to greater than 22G was and has been known to be a largely unpredictable phenomenon. The corrective actions applied per the Divergent Fuel Probe Oscillation THA were very effective at speeding the decay of the oscillation. Although unusual, the large oscillation did not cause structural damage to the airframe.

Even though the test team's safety planning accounted for the possibility of a large aerial refueling probe oscillation resulting in aircraft damage, the remainder of the test was cancelled. While the Air Force only required a minor safety amendment documenting the event and a minor delay to the test schedule, Sikorsky's Model Development Safety Committee deemed the event a mishap and prohibited the Sikorsky pilot and engineers from flying the remainder of the planned test. Since Sikorsky would not allow the Air Force to operate their test equipment, the remainder of the test could not be completed.

Lessons Learned

Numerous lessons were learned from the initial program introduction though test execution. Many corrections and process changes were implemented while the program was being executed rather than waiting for a future program. While some of the lessons could not have been anticipated, several other key areas were noted in advanced as being potential limiting factors and extra time and consideration were applied.

From the beginning, the joint execution of a test program with a contractor using a mixed DT and OT CTF was expected to be out of the ordinary and challenging. As a result, from the onset of the program, the test team developed a schedule that incorporated both military and contractor safety planning process. While the CTF knew what was needed from Sikorsky at certain times in the planning process to begin execution of the military led test on schedule (mishap responsibility and liability belong to the Air Force), the team did not realize that Sikorsky maintained internal policies to limit liability to customer aircraft, regardless of military regulations. While Sikorsky agreed that the military was the lead test organization, Sikorsky imposed additional safety restrictions outside the scope of the military safety planning that caused several last minute briefings and paperwork changes. The net result in the planning process was that of an iterative process, which caused problems and delays in the approval cycle. When separate follow-on testing was planned after the completion of this probe testing, the test team made sure to fully explain the approval process and the exact timing requirements to ensure on-time test execution.

Along the same lines as the conflict in safety processes, the team discovered that contractor technical test requirements were written less stringently and more flexibly than the military test typically allow. Numerous times during the planning process, Sikorsky added additional test equipment, test types, and test requirements that were not documented in the Sikorsky provided test plan, nor were provided in the appropriate timeframe for implementation. Because Sikorsky was responsible for data quality and insisted that the test would be incomplete or insufficient data would be collected without the changes, the Air Force test team had to re-accomplish some military related test planning and add additional equipment for SPO configuration control board approval.

The major lesson learned in working with a contractor was that they had a different approach to flight testing than the military, because they flight tested with a completely different purpose. The Sikorsky team was primarily concerned with acquiring valid and robust technical data on which they could make a causal determination for refueling probe oscillations. The military portion of the team was more concerned with taking a safety-centric approach to this program, based on the damage reports received from the operational units. The Air Force as a whole maintains focus on aircraft safety through the full range of testing from first flights to envelope expansion to operational testing in realistic scenarios. Again, the team learned that providing the contractor with a better understanding of military processes and intent when it came to developing technical and safety requirements would have helped. The test team also realized that a contractor's priority list is not the same as the military's priorities, even if the military was a customer. This test was the number one priority of the CSAR division of ACC; however, several other Sikorsky customers were deemed higher priorities. Many of the paperwork delays were caused by individual Sikorsky engineers being tasked to higher priority Sikorsky customer tasks. Understanding competing priorities and building in a buffer to due dates may help alleviate some schedule delays and frustrations.

Even though contractor flight on Air Force aircraft is common and use of chase aircraft in test occurs daily, mixed aircrew and test safety chase was relatively new to the CSAR CTF. The approval process, test related training requirements, differences between flight test in Major Commands (MAJCOMs, DT in AFMC versus OT in ACC), testing versus training requirements, and use of a GFR in association with this type of aircraft and test were not well known to the team. The team researched required regulations extensively in all areas and began their research at the initiation of the program. They approached the issue assuming that the regulations existed, but as it turned out, they did not fully exist. While several relevant regulations were found, several key requirements specific to the Air Force, the HH-60, DT, and contractors did not exist. The team did not consider that they might have to develop or supplement regulations for approval in addition to developing the safety planning for the test. While early initiation of research prevented severe delays in schedule, the team learned how long it can take to identify the correct approving authority, verify the regulations did not exist by the approving authority, then develop and coordinate appropriate regulations, and gain approval at the MAJCOM level. Test teams should identify anything "new" associated with their test and verify they will be able to test under existing regulations.

One of the biggest challenges of the test involved balancing technical requirements and safety requirements. The test equipment that was being used to collect all of the technically required data was also the same equipment that was used for safety abort and KIO calls. Because the military did not have its own independent equipment to monitor test safety parameters, and the contractor would not allow the military to operate their test equipment, the contractor had to monitor and make the test safety calls or the test would not happen. The test team actively discussed the ramifications of a MESP making these calls as opposed to a trained TC or pilot during the approval briefing, but the decision was made to allow the MESP to make the calls, considering the fact that the MESP was the technical expert of the data that was being monitored. The approval did require that a trained TC be onboard directing the test as a mitigation. A strict comm plan was developed and clearly defined calls and resulting actions were made. A dry run of several test scenarios was also executed on the ground prior to actual test execution. A familiarization flight was performed with all test aircrew and aircrew briefings were held. However, the differences between a contractor's style of testing and the military's style of testing were more profound than the team anticipated.

During various test points preceding the large oscillation that ended the test, the test team noticed that the Sikorsky MESPs focused on collecting technical data more than they were concerned about the safety parameters that they were monitoring. Their approach suggested that they were passengers passively collecting data, rather than critical aircrew actively monitoring for safety. Several times during testing, the Air Force TC made the abort calls for the Sikorsky MESP after observing the raw data on their screens exceeding the marked limits. The test team discussed these missed calls in post-flight debriefings and made a few interesting discoveries.

First, Sikorsky was visually performing a moving average on the raw accelerometer data to determine an abort call. While accelerometer data are inherently noisy and artificial spikes in acceleration levels are expected, the team realized that, since they assumed Sikorsky was the technical expert in structures and structures testing, they had inadvertently assumed that Sikorsky was monitoring these parameters for flight safety purposes. While the team knew exactly which accelerometer was being monitored, the team had not considered that Sikorsky's equipment was tailored for data quality, not safety monitoring.

The team also discovered that Sikorsky was using a separate measure (a Fast Fourier Transform of the signal) to determine if an abort had been reached within their post-test point review. As a result, the test team required Sikorsky to implement an actual moving average acceleration line on their screen and part of their screen to change color when limits were exceeded. One of Sikorsky's additional safety requirements imposed by their company's safety board involved a two-abort rule. If an abort was reached twice in one flight, the test had to be terminated for the day and a safety release had to be completed by their test personnel and approved by the company before resuming testing. Although the separate measure was more accurate from a technical perspective, Sikorsky appeared to use this value to determine if an abort limit had been exceeded because the value was almost always lower than the moving average, thus avoiding their abort rule. This use of an abort was fundamentally different than that used by aircrew and was different than the use briefed to the test team and aircrew. In this case, the test team determined that Sikorsky's abort requirements were more restrictive than those required through the approved safety process, and thus did not impose any restrictions on which parameters were used for this "Sikorsky Abort" determination. The discovery did highlight the difference in a safety approach versus technical approach to the test team. Rather than becoming concerned at the difference in approach, the team's confidence increased in Sikorsky's monitoring as the exact values of all the calls were now explicitly specified and all missed calls could now be explained. While the team thought that the issues had been resolved, the differences in test approach did not change, which had its greatest impact during the large limit-cycle oscillation.

During the limit-cycle oscillation, the Sikorsky MESP delayed his abort call to see if the oscillation was growing. That decision opposed the training of TCs, test crews, and Air Force testing philosophy: safe, effective, efficient test, in that order. After the delayed abort, the oscillation continued to grow. The Air Force TC realized the Sikorsky MESP was not monitoring for the Abort or KIO call and called an Abort as they had exceeded the Abort and subsequently KIO value. While it is difficult to determine the exact impact of the delayed Abort call and the missing KIO call, TCs and military aircrew are trained in identifying safety trends in test data. Since the difference between the abort value and KIO value was intended to be a buffer for small overshoots in acceleration due to the delay between the abort call and turning off the AFCS input, any growth in an oscillation should have resulted in an early KIO call. Additionally, the unusual nature of the oscillation compared to all previous test points should have indicated that an early KIO was warranted. While an appropriate call with a TC or pilot cannot be guaranteed, the test team believes the additional training and safety focused mindset of these aircrew make it much more likely. The biggest indicator of the difference in testing philosophy occurred the day following the oscillation event. Throughout the planning process, Sikorsky reiterated the fact that they were not trying to induce a large oscillation and only wanted to collect low-level vibration data. At the termination of testing, Sikorsky declared that they had all the data the needed to make the determination of root cause, even though only about 25% of planned testing had been executed, or marked as complete through analysis. The Sikorsky engineers stated that the limit-cycle oscillation provided the exact data they were looking for. The test team learned that while only minor aircraft damage occurred and no personnel were hurt during the test, something as small as a philosophy difference could have had a much larger impact. During separate follow-on probe testing, military aircrews maintained control of all safety related calls.

Even though safety is critical during testing, technical comm has to take place to make sure that appropriate data are being collected. With mixed civilian and military aircrew, determining the appropriate mix of technical and safety comms combined with periods of silence is challenging. From a technical perspective, more information is better; however, onboard the aircraft, the importance of each piece of information becomes diluted as the volume increases. At times during testing, the pilots found it hard to separate what sounded like important values from the technical discussions from the more critical safe to continue calls. The discussion-like nature of the post-test-point data reviews sounded more like questions on whether to proceed than necessarily a clear-cut good-data bad-data call. Because of the limited capabilities of the aircraft comm panel and the potential of contractor crew forgetting to switch the intercom to the entire crew when making safety calls, the entire aircrew completed all internal comms on the same channel. While the team had limited technology options, the team learned that having an even more rigorous comm plan with a directive action coming from one key person to the PF would have helped comms be more clear. While safety comms always have priority over technical comms, the ability for critical safety calls to be sent and discerned over all other communication is imperative.

One of the big successes of the large limit-cycle oscillation event was the safe aircraft recovery execution and following emergency landing at a local airfield. A large amount of planning was placed into the "what-if" scenario of a large oscillation. Because structural testing was taking place, a chase aircraft was planned for the test and was central to the corrective action strategy. Aircrew for both aircraft spent extensive time planning and practicing aircraft placement for a BDA, test monitoring, transitions between lead and trail, and entering and exiting airspace. The crew planned safe separations and other recovery procedures. Comms between the aircraft and the transitions in an emergency were planned. Multiple field landing points in the airspace along with local airfields were visited in advance and marked as waypoints in their navigation systems. A total of four sorties using two aircraft were flown for training and familiarity in preparation for the test. A quick reaction safety reporting checklist incorporating three Wing Safety offices across three bases was created and was practiced using a table-top discussion. As a result of this intensive planning, the test team was fully prepared and successfully executed the emergency planning. Both aircraft safely separated during the oscillation, a BDA was performed, the chase aircraft took lead and led the test aircraft to a safe landing site at a local airfield, and both aircraft landed safely, in spite of test aircraft damage. Safety notifications were submitted in under two hours and return of the aircraft to base occurred within a few hours. While no test team expects the worst will happen, the team learned that the time and effort in the planning process is well worth it.

Even though test teams spend significant time mitigating the worst case scenarios, the importance of the mindset that the worst could happen at any time can sometimes be forgotten. Efforts to account for all possible events can artificially raise the sense of safety, especially when a test can be moved into a lower risk category due to sufficient mitigation. One of the important lessons the team learned (one that all aircrew need reminding from time to time) is to never let past results influence readiness for the next test point. Testing at the CSAR CTF leading up to the probe oscillation testing had mostly been benign systems type testing with little continuity with previous, more complicated, and risky testing. The CSAR culture also relies on strong aircrew experience in operational scenarios to respond to threats and problems. These factors, combined with lack of experience in structural testing and the rapid onset problems that can occur when instabilities and non-linearities are encountered, made the test seem typical, predictable, and manageable. The team learned that they could have prepared more thoroughly for the test by studying previous lessons learned from other structural testing, adapting mission planning, briefing and debriefing methodology to enforce the uniqueness and unpredictability of the test, and could have studied techniques from other comm response critical testing such as flutter and ground collision avoidance testing.

Another almost trivial, yet potentially expensive lesson learned occurred the day prior to the beginning of the test. The test team had loaded a large quantity of contractor test equipment into the back of a truck for transportation to the aircraft the following day. When the contractor was asked if the equipment could get wet, they said it could, because it was used regularly in helicopters that regularly leak when it rains. The test team interpreted that answer to mean the equipment was waterproof and it was ok to leave it outside. While Las Vegas, Nevada typically has 294 days of sun per year, 22 days of measureable rain, and only a 20% chance of rain overnight, Nellis AFB experienced the biggest downpour and flooding event in several years that night. The next morning, all of the equipment, including the test critical \$100,000, one of a kind, data processing CPU was waterlogged. Sikorsky dried out their computers in a hotel room oven for the next 48 hours to salvage the equipment. Luckily, there was no equipment damage after everything dried out. The test team learned that the extra precaution of a tarp or storage of equipment indoors for critical equipment is a very minor task compared to the potential impact of not doing so. While it seemed like one tiny detail in comparison to the other massive logistics taking place, the impact should have warranted better attention.

Conclusions

While executing the first structural flight testing on the HH-60G in several years, making it the first structures test for the majority of the test team, the CSAR CTF faced several challenges that led to lessons learned, or re-learned, that are applicable to flight test regardless of test platform. The first, and probably largest lesson learned, was the importance of understanding differences in test approach between government and contractor organizations. Having an open and frank discussion on test requirements, both for process and data, could have helped this program run smoothly and limited the impact of a couple other lessons learned.

For example, the difference in approach to this test led to some of the in-aircraft communication concerns of the CTF. Due to the contractor's need to get data to determine the oscillation's cause, they allowed test points to extend beyond the pre-briefed Abort limit on occasion. That mentality led to a delayed Abort and KIO call on the point that ended in a limit-cycle oscillation. Having a military or government test conductor leading test safety calls may have helped.

In addition, initial testing led the team to believe that they could adequately predict the magnitude of response to a signal inject, which then led to an attitude that the aircraft and probe response during this phase of testing would be benign. During flight test in general, the test team must focus on expecting the unexpected. In this case, knowing that probe oscillations in the past had been so severe that a probe departed the helicopter should have keyed the team in that a severe oscillation was possible. Flight test is risky; teams should take time to review past testing and lessons learned and anticipate the potential outcomes of the test.

One of the successes of the test program was in the preparation the team put in case there was a safety event. Developing and practicing a build-up training plan and safety quick reaction checklist in advanced, led to a smooth recovery during the limit-cycle oscillation event with no further incident. Have a plan ready and standing by; hopefully there will not be a need to exercise it, but after a safety event is not the time to be developing a response plan.

Finally, while it seems somewhat trivial, the potential loss of the contractor's one-of-a-kind test equipment would have crippled this test program. Even a small detail in the greater scheme of a test program could turn out to be a make or break issue when the time comes.

Aircraft Controllability and Primary Flight Displays - The Baby is Ugly

Author Knut Lande (AF)

LandAvia Ltd, NO-4052 Royneberg, Norway
www.landavia.no
e-mail: knut.lande@landavia.no

ABSTRACT

LOC-I accidents are occurring every year, and the recommendations from the accident investigation reports seem to have no effect. Up to now the Accident Reports do not seem to address the reasons for pilots losing control, other than focusing on need for better or more pilot training. Little or no focus has been directed towards why pilots are losing control. During last year's ISASI conference in Augsburg a paper was presented discussing *"A New Tool for Analysing the Potential Influence of Vestibular Illusions"*[1], like Somatogravic and Somatogyral illusions resulting in pilot spatial disorientation. The knowledge that the human brain is depending on strong visual cues for orientation and balance in a zero or offset gravity environment is well documented.

However, this knowledge seem to be absent in most LOC-I accident reports. During flight conditions in darkness or in instrument conditions pilots may be affected by Somatogravic, Somatogyral and G-excess effect illusions resulting in pilot Spatial Disorientation ("pilot vertigo"). The only effective cues to combat these sensory illusions are strong visual cues. An illustrating point in this respect is the example of an untrained private pilot who enters clouds. It will not take long until he loses control, but chances are that he will recover control when the aircraft exits the cloud cover and the pilot observes the natural horizon line in daylight visual conditions. In this situation the untrained pilot receives visual cues from the natural horizon line filling his windscreen and by use of his peripheral eye sight he receives the required visual cues to combat the sensory illusions and restores aircraft control. This is an indication of the importance of a large and human factor centric artificial horizon on the pilot's Primary Flight Display.

During the last 25 years there have been investigated more than 18 LOC-I accidents and incidents globally where spatial disorientation may have been a factor. Still this was not a major focus by the accident investigations. This is an indication that the root cause of these accidents has not been identified. In none of these accidents did the accident investigation reports recommend improved Primary Flight Displays. The Air Asia QZ8501 is an example.

Several LOC accidents are related to aircraft stall. This may be an indication that the state-of-the-art Primary Flight Displays (PFD) are not presenting the aircraft flight condition in a most intuitive way, and not presenting the pilot with a clear indication of Angle of Attack (AOA). The Air France AF447 is an example in this respect.

---

[1] Reference 28

Using today's technology, flight displays could be made more intuitive and easier to read in an unexpected and blurred flight situation. Also, by making flight control inceptors and autopilot modes more intuitive, and thus improve tactile feedback cues, the pilots may be kept in the loop and be more prepared to take control in an unexpected situation.

Traditionally the test pilot community has focused on aircraft handling qualities. The result is almost perfect handling qualities in modern aircraft. Regardless of how ideal the aircraft's handling qualities are, it is of limited value if the handling pilot becomes confused and disorientated during an unusual attitude (UA) or upset recovery (UR). Several of the recent accidents seem to be related to PFD's and the pilots knowledge and training in interpreting their displays. PFD is vital for the pilot's ability to keep Situational Awareness (SA) and it may be time for the manufacturers and test pilots to focus on PFD's in addition to autopilot modes and aircraft handling qualities during test and evaluations.

The link between the pilots and the aircraft – the man-machine interface – should be improved and made more intuitive. Today's technology makes it possible to improve the displays and autopilot modes in aircraft based on Human Factor research.

INTRODUCTION

The global air accident statistics have gradually improved over the last 50 years. This has mainly been caused by gradual improvements in aeronautical engineering and technological innovations. Examples on this are improved Weather Radars (WR), Navigation Displays (ND), Vertical Displays (VD), Traffic and Collision Avoidance Systems/Airborne Collision Avoidance System (TCAS/ACAS), Ground Proximity Warning System/Terrain Avoidance and Warning System/Enhanced Ground Proximity Warning System (GPWS/TAWS/EGPWS), Category II/III Landing Systems, etc.

The latest technological advancements helped reduce the CFIT categories of accidents. However, the reduction in CFIT accidents led to increased focus on Loss of Control in Flight (LOC-I) accident statistics which seems to remain high.

ACCIDENTS

Typical LOC-I accidents

The common factors in most LOC accidents seem to be pilot confusions during unfamiliar flight situations, lack of pilot knowledge and piloting skills. From the CAST Team report in Figure 1 we see that most LOC-I fatal accidents are related to attitude and AOA control.

We see that in most LOC-I type of accident reports, the focus is on piloting knowledge and skills. No reference is made to possible loss of spatial orientation contributed by less than optimal Primary Flight Display (PFD) or lack of Angle of Attack indication. It is the author's opinion that the progress in modern Glass Cockpit and Flight Display design have improved the navigation task, but not contributed to improving pilots' attitude in space situational awareness. It may be argued that some of the modern Flight Displays are not very intuitive or helpful in a loss of control situation.

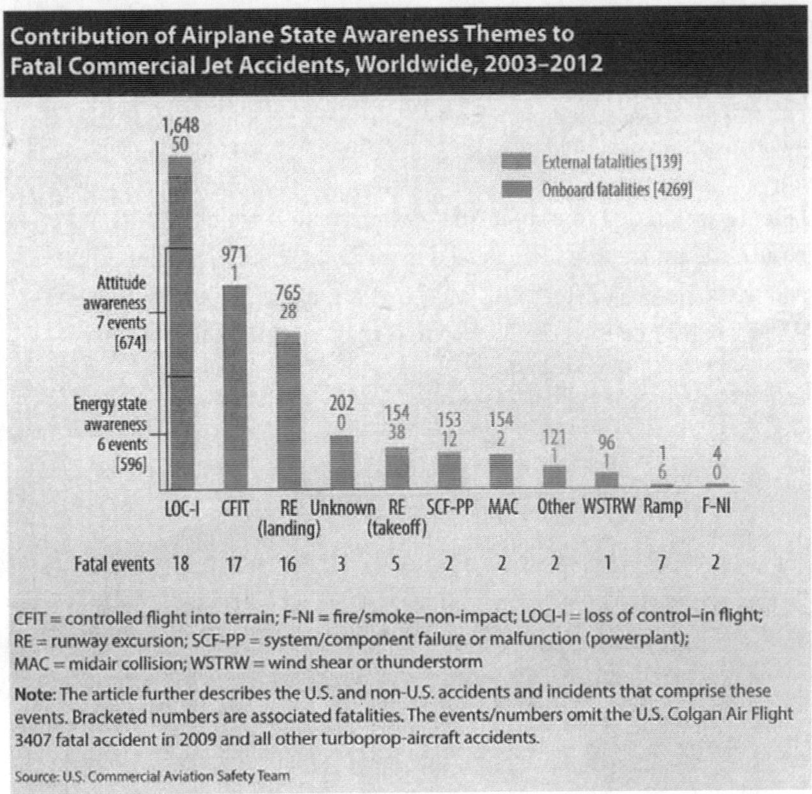

Figure 1. Commercial jet accidents, worldwide 2003-2012 (US CAST).

Further, the increased use of cockpit automation seems to have influenced the education and training of younger pilots. We have also seen that manufacturers advertise their airplanes as easy to fly (*"low workload"*) and *"unable to stall"*. Such statements are unrealistic and misleading to student pilots. Any student pilot should have a basic knowledge in physics and be taught from basic training that any heavier than air vehicle must always be aerodynamically controlled and may *"fall out of the sky"* if the flight conditions producing a sustainable lift are not maintained, i.e. a minimum airspeed, or more accurately, an Angle of Attack (AOA) below maximum allowable (below the stall AOA). Whether the airplane is stalling or out of control for other reasons does not matter.

Even though todays accident rates in aviation are quite low, the latest LOC-I accidents are avoidable and seem to be caused by a combination of inefficient PFD's and lack of knowledge and basic flying skills among the pilots. We find that LOC-I accidents results from stalls both during high altitude cruise and during approach.

These types of accidents are also happening to helicopters, as seen from the accident to a North Sea helicopter which crashed off Shetland. The helicopter lost airspeed and entered wortex ring state which may be comparable to a fixed wing stall.

In a typical LOC-I stall accident the pilots did not recognize the slow speed/high alpha flight condition and were not able to prevent the accidents. We see that these types of accidents are not only related to fixed wing aircraft. While airplanes may stall and enter controllable deep stall, helicopters may enter controllable Vortex Ring State (VRS) which, from a piloting point of view, may be comparable to an airplane stall. The recovery is similar - nose down to break the condition and increase power. However, a safe escape requires altitude just like an airplane stall at low altitude.

Other types of LOC-I accidents are related to pilot spatial disorientation which results in pilot confusion and loss of attitude awareness. The spatial disorientation may be caused by Somatogravic, Somatogyral and G-excess Effect illusions[2] causing the pilots to become spatial disoriented, and pitch or bank too much or in the wrong direction. Such mistakes may cause loss of control.

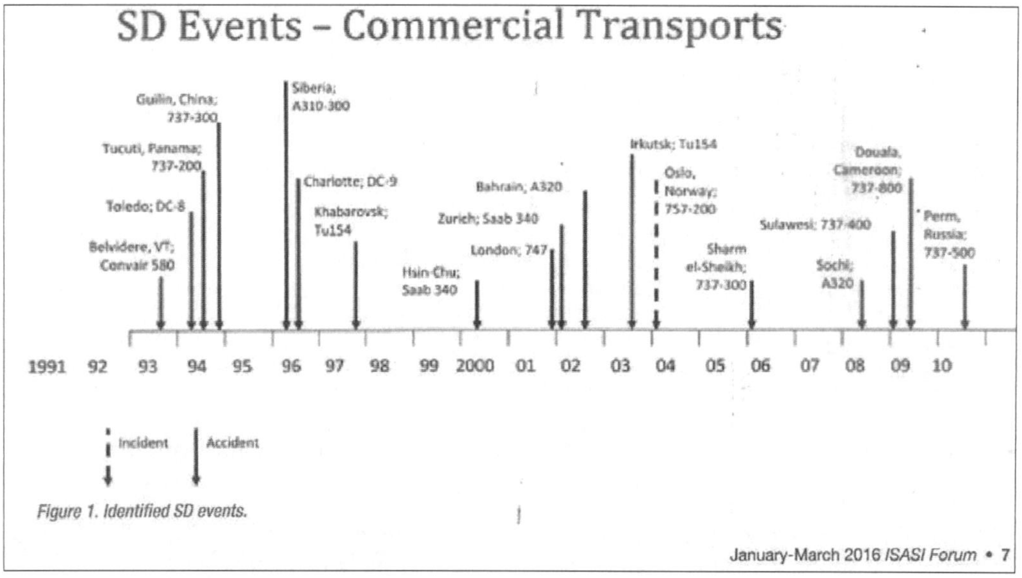

Figure 2. Spatial Disorientation in Commercial Transports (Dr. R. Mumaw et.al 2016). [3]

Accident Investigations

Historically, most LOC accidents were labeled «Pilot/Human Error».

Professor Dr. Sidney Dekker offers a "new view" on Human Error (2006)[4]:
The Old View: "*Human error is a cause of trouble*" (Bad Apple Theory).
The New View: "*Human error is a symptom of trouble deeper inside a system*".

Professor Dr. James Reason (1997)[5]:
"The Organizational model views human error more as a consequence than as a cause. Errors are the symptoms that reveal the presence of latent conditions in the system at large".

In modern accident investigation theory Human Error is not considered a cause of accidents. LOC-I accidents have several underlying cause factors and most LOC-I accidents are

---

2 Reference 27
3 Reference 28
4 Reference 14
5 Reference 15

Organizational Accidents.

Figure 3 shows the typical operational and training flight envelope of today's commercial pilots. How can we expect pilots to cope with a flight upset situation involving more than 90° of bank, when they have only seen 60° during their training and are normally limited to 30° during normal operations?

Clearly then, we may label a LOC-I accident as an Organizational Accident and not just Human Error.

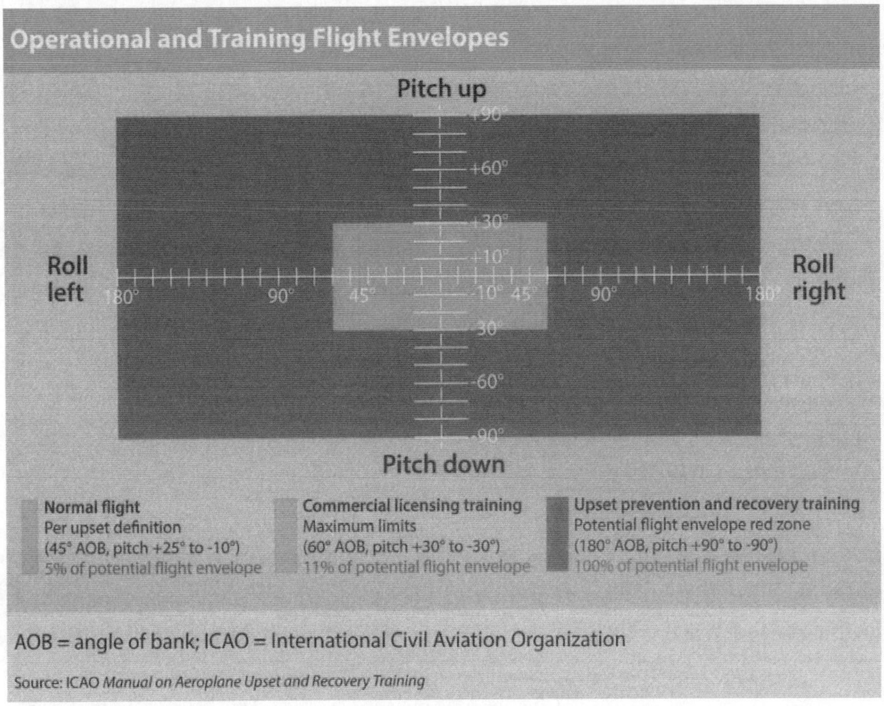

Figure 3. Operational and Training Flight Envelope (ICAO).

AREAS OF CONCERN

Coffin Corner

When flying at high altitude the air density is low; hence lift and thrust are reduced. The airplane angle of attack is increased to compensate for the loss of lift. The mach number[6] increases as a function of lower temperature and lower speed of sound. Thus the maximum cruise true airspeed (VMO) approaches the critical mach number (MMO) and must be reduced (mach buffet). Similarly, the true stall speed/mach increases. As the airplane climbs the maximum cruise TAS is gradually reduced and approaches the minimum flying speed or critical angle of attack (stall buffet). Hence, the margin between high mach buffeting and low speed buffeting becomes smaller and smaller with altitude. This flight conditions has been labeled the "*coffin corner*" due to the challenging flight conditions, Figure 4.
Hand flying at high altitude by reference to airspeed only is very challenging. It is very difficult to control the airspeed accurately and angle of attack indication is required for better

[6] Mach number is the ratio TAS/Speed of Sound

airspeed control. Civilian trained pilots of today have not had basic training in this area ("pitch and power") and have limited hand flying skills.

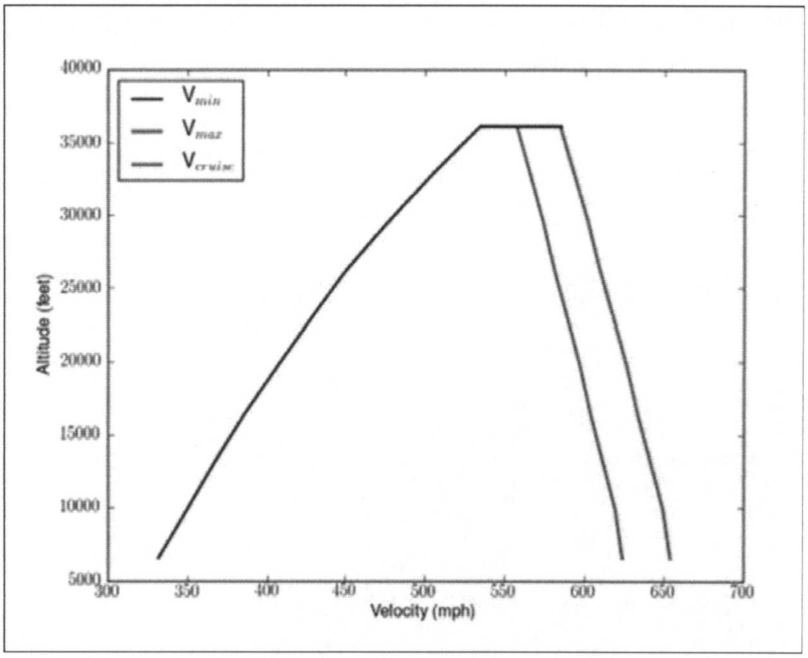

Figure 4. Coffin Corner (Wikipedia 2012).

Aerodynamic Stall and lack of AOA indicators

Aeronautical knowledge about the relationship between angle of attack and flying speed, and the necessary training has been basic pilot knowledge and training requirement for decades. However, based on the recent LOC-I accidents it seems that the focus on stall theory and training has diminished.

Aerodynamic stall occurs when the airspeed reduces below stall speed which varies according to flight conditions, Figure 5. However, it is not the airspeed itself but the angle of attack which is the critical factor, Figure 6. The airspeed is an indirect indication of the angle of attack in level flight. During maneuvering flight the wing will still stall at the $C_{L\ max}$ but at a higher indicated stall speed[7]. Hence, it is only in level flight the pilots have a good reference to the actual stall speed. An angle of attack indicator will give direct indication in relation to the stall angle of attack. This is basic pilot knowledge, but for some unknown reason the aircraft manufacturers and certifying authorities have been reluctant to include an angle of attack indicator as a primary flight instrument.

[7] $V_{S2} = V_{S1}\sqrt{n}$

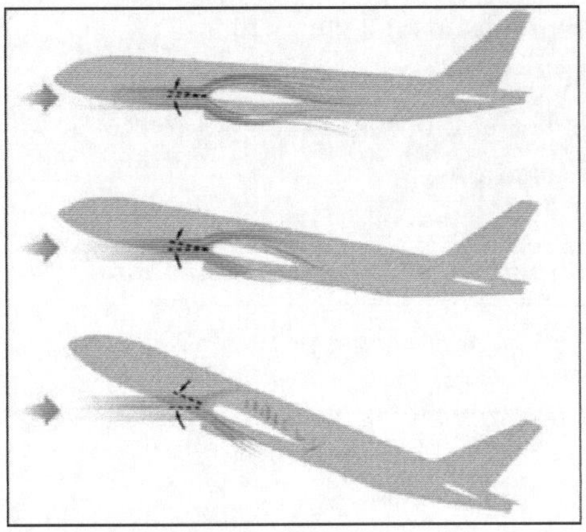

Figure 5. Aerodynamic Stall
(D. Carbaugh, Boeing 2010).

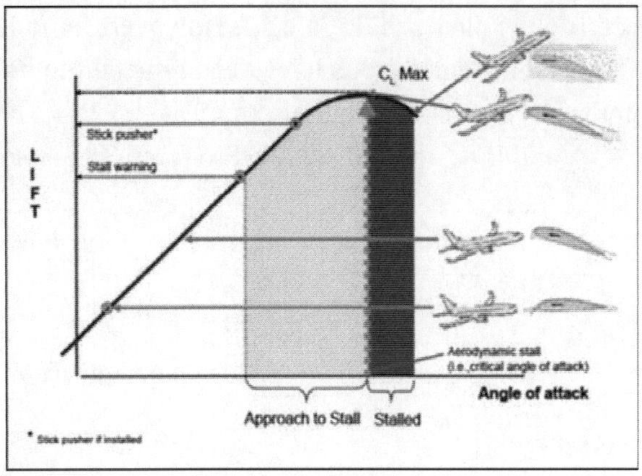

Figure 6. Airplane stall angle of attack
(D. McKenney, ALPAI 2010).

Stalls can occur when performing any maneuver. The wing "does not know" about airplane attitude or airspeed, Figure 7.

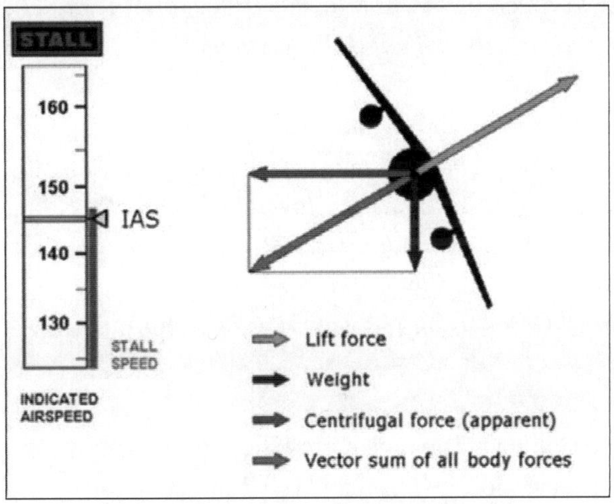

Figure 7. Stall during maneuvering (Wikipedia 2012).

The wing "stops flying" when the critical angle of attack is exceeded. Result is a stall - and if not properly educated and trained, the pilot may lose control. An Angle of Attack (AOA) indicator which clearly and intuitively tells the pilot the margin to stall or minimum control may mitigate such loss of control. This is the most fundamental and basic knowledge the student pilots are required to know, and that the airspeed indicator is just an "aerodynamic indicator" which does not tell the pilot how much lift margin he has, but is related to the forces acting on the aircraft. This has been an item for discussion for several decades. Technically it is not complicated to install useable and intuitive angle of attack indicators in airplanes. This is merely a policy issue. US Navy and Air Force fighter airplanes have had alpha indicators installed for years, in addition to stall warning and artificial stall recovery systems.

Pilots used to alpha indicators found these very intuitive and useful, showing the lift and performance margin in maneuvering and accelerated flight as well in approach and slow speed flight conditions, Figure 14. However, for unknown reasons it appears that FAA, EASA, Boeing and Airbus have been reluctant to provide pilots with this information. Reference 18 describes flight in the SR-71A Blackbird. The author makes several references to AOA and instrument displays. From ref. 18 are some citations:

> *"Practically all high-speed military jets have an angle of attack (AOA) indicator in the cockpit."*
>
> *"AOA is an excellent indication of how well (or how poorly) an airplane is flying through the air."*
>
> *"Until AOA indicators were developed, pilots had to know their aircraft's flying characteristics at all gross weights, fuel loads, g-loads, flap settings, airspeeds, and other variables that affected the plane's stall speed. An AOA indicator took all those variables into account and became a reliable indication of how well the aircraft was performing. In high-speed fighter aircraft, this development was a tremendous improvement because the pilot had only to check his AOA indicator to see how his aircraft is performing. Once an aircraft reached its critical AOA, it stalled."*
>
> *"When I applied for the SR-71 program in 1973, part of the evaluation process at Beal consisted of two T-38 rides with an experienced SR-71 pilot making sure your flying skills and general airmanship were good. I had never flown the T-38 before and was somewhat apprehensive about how the plane handled. My acceptance into the SR-71 program was riding on how well I could fly it. I distinctly remember my evaluator telling me to fly the T-38 just like it was an F-4 and to use the AOA. Once he said that, I flew it easily."*

As a result of the many LOC-I accidents in recent years, there is a growing interest in installing AOA indicators as part of the PFD. This author, just as many other military trained pilots of the "cold war era", has flown several aircraft with AOA indicators. The experience gained is that AOA is a very intuitive primary flight indicator and just as valuable as the airspeed indicator.

The author has read several comments from pilots not used to AOA indicators, supplemented by some test pilots caution against reliance on AOA indicators, with reference to inaccuracies and limitations. AOA is measured on most modern aircraft and the data is used in the aircraft computers to calculate various flight parameters. Further, the AOA indication is of most importance when maneuvering or flying close to the stall AOA. Hence, the indication will be most useful during slow speed flight or during LOC and upset recoveries. However, the AOA must be a prominent and intuitive indication based on Human Factor and Aircraft Flight Mechanics theory which is taught in basic pilot training.

Boeing is offering AOA indicator as an option, Figure 8, but this is not the most efficient type of display.

Figure 8. Boeing AOA indicator (Boeing 2013).

DEVELOPMENTS IN FLIGHT DISPLAYS

LOC-I is not only related to pilot knowledge and skill, but just as much to Flight Displays, Control Sticks and Levers (inceptors/thrust levers), in other words the Human – Machine – Interface (HMI).

It is worth noticing that in the early days the pilots did not have a stand by attitude indicator available. To compensate the military pilots trained Partial Panel where the Attitude Indicator was covered or disconnected.

Figure 9. Lockheed/Canadair CF-104 (Photo: Marc Bourque via K. Lande).

With the introduction of gyro stabilized inertial platforms the stand by attitude indicator became standard, but there was still no instrument layout standardization, Figure 9. Note the large instrument sizes which made situational awareness in unusual attitudes easier. It is also worth mentioning the large Attitude Indicator (AI) with heading indication. With reference to just one indicator one got an instant "3D sense" of the flight attitude. This type of AI was used in several US aircraft, Apollo and Space Shuttle during the 1960's and 1970's. Compare this with today's electronic ADI which does not give the pilot the same "3D" sensation of "attitude in space", but only gives a "flat 2D" impression of attitude.

FAR 25 specified the flight instruments arranged in a T-shaped pattern, with the ADI in center, Horizontal Situation Indicator (HSI) below, airspeed located to the left of the ADI, altimeter to the right, with the vertical speed immediately below the altimeter. This arrangement became known as the "*Basic–T*".

Figure 10. Commercial Airliner (1970's) (K. Lande).

Figures 10 shows the FAA standardized flight deck of a commercial airliner of the 1970's. This standard "*basic–T*"[8] lay out is still a certification requirement for modern aircraft in FAR 25 and EASA CS-25, Figure 11.

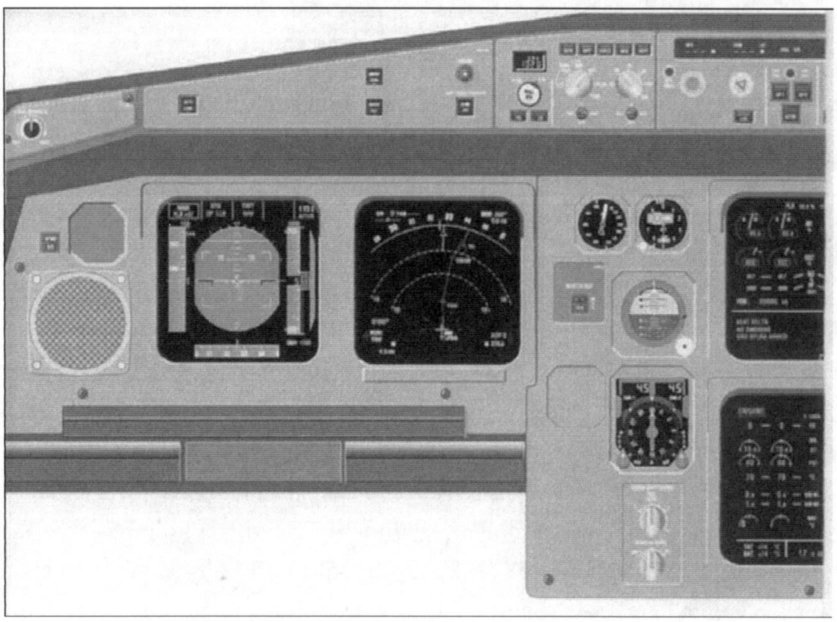

Figure 11. Airbus standard flight display layout (Airbus).

[8] Reference 13

We see that the relationship between the ADI and the other flight instruments are retained in the "*basic–T*" fashion. However, it may be argued that even if the scan distance is reduced, so is the instrument readability.

Another aspect is the regulated installation of stand by instruments but no associated requirement for training in use of these instruments[9].

Figure 12. Conventional airspeed indicator (Wikipedia 2014).

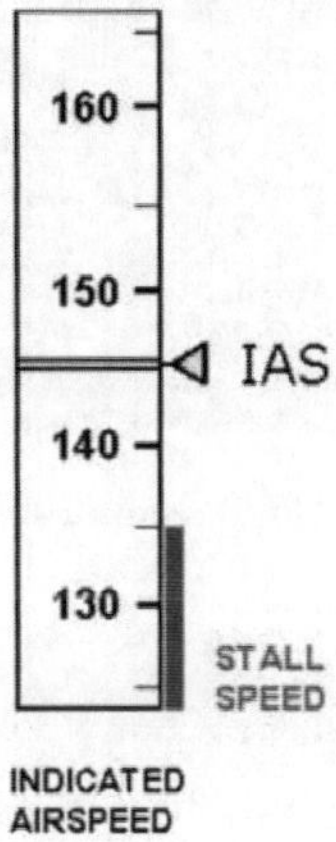

Figure 13. State of the art airspeed indicator (Wikipedia 2014).

In an unusual attitude situation the old fashion round dial airspeed indicator is easy to read, with large numbers, solid white needle pointer and a intuitive indication, Figure 12. With the pointer in the slow speed region on the right hand side of the airspeed indicator, a pull on the stick will increase the angle of attack and reduce the airspeed. The approximate position of the pointer in relation to the position and color on the scale give the pilot an approximate sense of the airspeed and the lift margin at a glance, in addition to giving the pilot a good rate of change in airspeed.

The same type of instrument could just as well display AOA, Figure 14. The instrument is intuitive in the sense that pulling on the control stick/yoke increases/rotates the indicator needle to a higher digit and vice versa (pull-increases AOA, push-decreases AOA). What can be more intuitive to a pilot than pulling on the stick that increases the AOA/reduces the IAS, and pushing on the stick that decreases AOA/increases IAS.

The human brain is analogue and there seem to be human factor indications that analogue round dial instruments may be easier and quicker to read/interpret than digital indications . This may be important during dynamic flight situations. A reminder to this effect is that most people use wrist watches with round dials and pointers (analogue round dial display) in lieu of watches with digits (digital display). Further, most aerobatic pilots prefer large round dial instruments for quicker scan.

---

9 Reference 8

Figure 14. AOA indicator (Wikipedia 2014).

Figure 15. Typical analogue wrist watch (K. Lande 2016).

Figure 15 show a typical wrist watch without numbers. The round dial analogue airspeed and AOA scales may be compared to a typical analogue wrist watch. It is so intuitive that numbers are not required. The user may see the approximate time at a glance. If more accurate time reading is required, a closer look may be required. This priciple works just as fine for aircraft PFD in a cockpit.

It is interesting to note the format developed by USAF in 1958, and introduced in operational aircraft such as F-105, F-106, F-111, C-141 and C-5, Figure 16. USAF research concluded that the airspeed scale digits should be increaseing downwards, i.e. "lower the nose, increase the airspeed" ("fly-to-principle").[10]

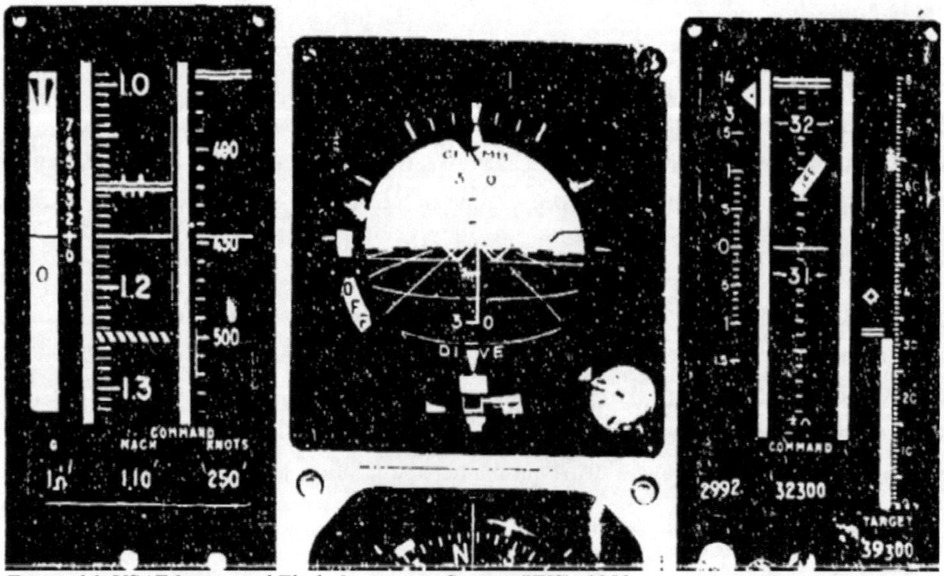

Figure 16. USAF Integrated Flight Instrument System (IFIS), 1958.

[10] Reference 3

On the other hand, the present state-of-the-art airspeed indicator is not intuitive, even if they have a “speed trend indicator”. This is not optimal as a rate indicator, especially in turbulence. The scale numbers are increasing upwards. This is opposite to raising the aircraft nose (pulling on the stick). From basic training every pilot is taught that raising the nose will decrease the speed and lowering the nose will increase the airspeed.

Further, by raising the nose and hence reducing the airspeed, the moving scale should move downwards with decreasing digits from the top[11]. This is in line with Human Factor spesialist recommendations. In its present format the state-of-the-art airspeed indicator is not intuitive and the vertical scale should be reveresed. Pushing the aircraft nose down would then result in increasing airspeed. This is also recommended by Don Bateman of Honeywell.[12]

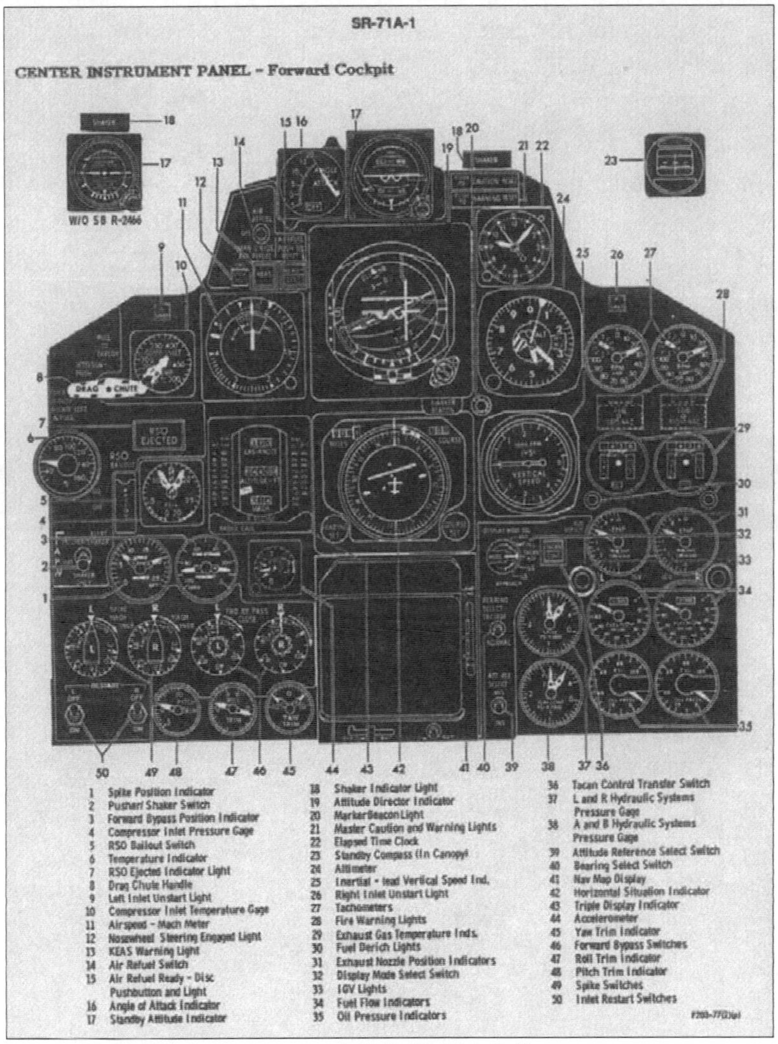

Figure 17. SR-71A Instrument panel (T.O. SR-71A-1).[13]

[11] References 9

[12] Reference 21

[13] Reference 18

Figure 17 show the instrument panel of the SR-71A. This was a very challanging aircraft to fly and demanded good pilot knowledge and flying skills. It was an aircraft which needed continous pilot attention to attitude and AOA. This was a USAF aircraft with an instrument panel developed concurrently with the previous mentioned IFIS. Still we see a conventional "Basic-T" layout with round diales. We see that a large portion of the instrument panel area was devoted to the primary flight instruments. Compare this to Figure 11 which is a state-of the-art primary flight display. We also see the prominent location of the AOA indicator close to attitude and airspeed. The large round dials with prominet pointers allowed easy reading and permitted use of the peripheral vision registering the relative position of the pointer. It allowed the pilot to interpret the flight condition without necessarily reading the exact number.

Initially, SR-71 pilots were sceptical to night flying, and there were some incidents where pilots overcontrolled in bank during turning at Mach 3. This was caused by a sensory illusion called G-Excess Effect, Figure 18. To compensate for this Lockheed developed a device called Peripheral Vision Display (PVD). It projected a laser generated, thinly red line parallel to the horizon, across the pilot's instrument panel. The PVD was not intended to be part of the pilot's instrument crosscheck. Instead, the pilot perceived the laser line indirectly by peripheral vision and subsconciously supported spatial orientation, just as visible outside horizon supports orientation during daytime flying. As the aircraft pitched and rolled, the red horizon line also pitched and rolled across the instrument panel and gave the pilot instant orientation. The horizon line flashed to warn the pilot if the pitch or bank angles exceeded certain limits.

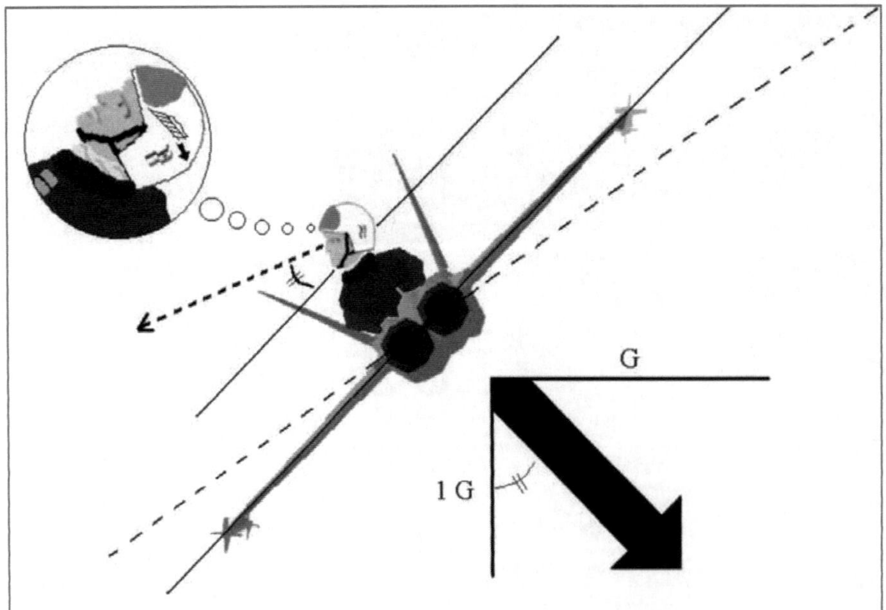

Figure 18. G-Excess Effect during turns (Dr. Bob Cheung, (DCIEM) Canada.[14]

[14] Reference 27

The digits on state-of-the-art flight instruments are also smaller than previously and not so easy to read in a "blurred" and dynamic flight (LOC) situation. The same may be said about the altimeter and vertical speed scales. Such scales are more suited to computers than to aircraft displays. These types of displays were introduced during the 1980's and were driven by the avionics manufacturers. I have seen limited research reports documenting the benefits and efficiency of the modern primary flight display indicators, other than engineering and cost benefits[15]. According to available litterature regarding vertical scale instruments and movement, the benefits of todays flight displays in a LOC situation is doubtful[16]. I am convinced that modern glass flight displays have improved the horisontal navigation and helped reduce CFIT accidents. However, they are less intuitive and efficient in resolving the pilots sense of the "3D attitute in space" in a LOC situation (unusual attitude or upset situation), and may have contributed to several of the recent LOC-I accidents. With todays digital technology it should be possible to design more functional and intuitive Flight Displays.

The flight displays in most commercial aircraft of today were introduced during the 1980's and have not changed much, even though todays digital technology make most designs possible, Figure 19. Figures 20 and 21 show state-of-the-art flight decks which will be around for the next 20-30 years. The layout is basically the same as developed 30 years ago, and is also, with a few exceptions, standard in modern helicopter cockpits. Considering the number of LOC-I accidents worldwide, it may be time to reconsider the present PFD layout and do some HMI research and development of flight displays and controls. It may be time to look for new types of displays which are more intuitive, based on human factor considerations and basic flight mechanics principles.

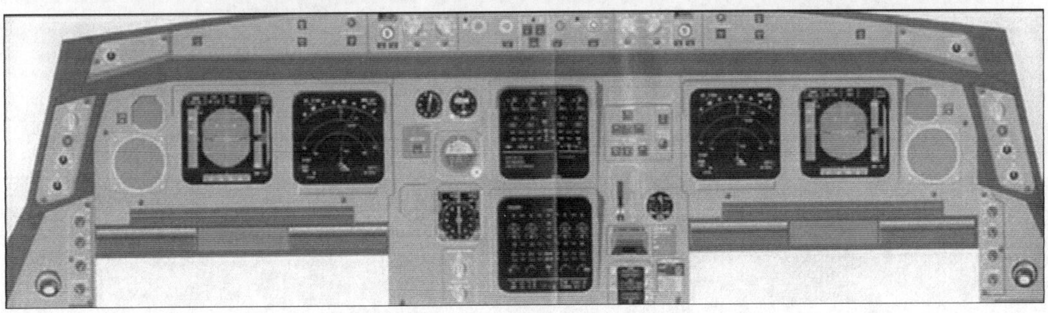

Figure 19. Airbus 320 Flight Deck (Airbus1982).

The author has several hundred flight hours with similar types of displays, but was not introduced to any recommended scan pattern. The general idea was that all the information is included on one display, "right in front of you". However, several LOC-I accidents in recent years may indicate that the state-of-the-art PFD's are not as efficient as thought of in the industry.

[15] References 7 and 9

[16] References 4, 5, 7 and 9

Figure 20. Airbus Flight Deck (Airbus 2012).

Figure 21. Boeing Flight Deck (Boeing 2015).

Figure 22. Modern Flight Displays (Internet 2016).

These types of PFD's are satisfactory only during benign flight conditions and autopilot operations, but not during a dynamic or upset flight situation leading to an unusual attitude. In such flight conditions all flight parameters are changing and the human brain may be saturated by changing parameters and digits that need direct readings and interpretation. Further, todays PFD's become more and more cluttered with non-flight-control-essential indications which may tend to distract the pilot's attention.

Today's state-of-the-art instrument panel seems to have lost some of the benefits of the older "classic" panel. The panel seems very clean and is certainly saving space (and money). It is quite sufficient during routine benign flight conditions within 15° of pitch and 30° of roll attitude, but not optimal during a dynamic unusual attitude (LOC-I) recovery situation.
It seems that the larger the screens the more data are included, making the PFD very cluttered.

A large screen with wide horizon line from the synthetic picture of the terrain and transparent round dials with pointers, including AOA indicator, will enhance the pilot's situational awareness.
Being an experienced flight and aerobatic instructor the author has seen the reactions of students with limited or no outside references. During such flight conditions the student pilots loose aircraft control very quickly. However, once visual reference to the outside terrain and horizon is regained, they may regain aircraft control within reasonable time and altitude loss. It follows that many of the recent LOC-I accidents might have been avoided if the pilots had visual sight of the underlying terrain.

Therefore, future Primary Flight Displays should be based on a synthetic picture of the outside world with overlaid prominent and transparent primary round dial flight instruments, including a prominent AOA indicator. Figure 22 shows two later PFD displays for fixed wing aircraft.

AAIB[17] investigated a LOC-I incident that happened to a single pilot aircraft which stalled at high altitude in 2013. Quote from the AAIB report[18]:

"The pilot then recalled a violent and very confusing rolling departure from controlled flight. The aircraft almost immediately entered high cirrus cloud, obscuring the horizon. The pilot was unable to interpret the PFD attitude indicator, which he described as presenting information that he could not recall having seen before. He did not recall exactly how long this persisted but he did recall checking both the left and right PFD displays, which were similar in appearance.
The pilot made several attempts to recover the aircraft, although he could not later recall what control inputs he made. He recalled selecting idle thrust and achieving almost level flight at one point but, he had not increased thrust and the aircraft slowed rapidly and again departed from controlled flight. During the period that the aircraft was out of control it descended into clearer air between cloud layers, with a visible external horizon allowing the pilot to regain control."

In most LOC-I accidents the pilots do not survive to tell their story. Here we have an example of a pilot who stalled his aircraft and was not able to interpret his state-of-the art PFD. It is clear that the pilot was totally confused, and only regained control when he could refer to the natural horizon. He survived and could tell the investigators his experience.

Figure 23 shows an Advanced Flight Systems (AFS) type of PFD installed in some GA airplanes. The positive feature is the display of the terrain and transparent flight instruments, but the negative features are the strip indicators and all the added information making the PFD cluttered.

In 2010 NTSB issued a Study Report on accidents in GA airplanes with Glass Cockpit compared to airplanes with conventional flight instruments. The report shows that accidents in the 2002-2008 periods resulting in fatalities were 16% with conventional flight instruments, compared to 31% with Glass Cockpit.
On average, glass cockpit aircraft had:

- More accidents during climb, cruise, and approach
- More loss of control in flight, collision with terrain, and weather encounters
- More personal/business flights
- Longer flights
- More IFR
- Accident pilots were older, with more flight hours and higher certificates/ratings.

---

17 Aircraft Accident Investigation Branch (UK)

18 AAIB Bulletin: 1/2015 N380CR EW/C2013/12/05

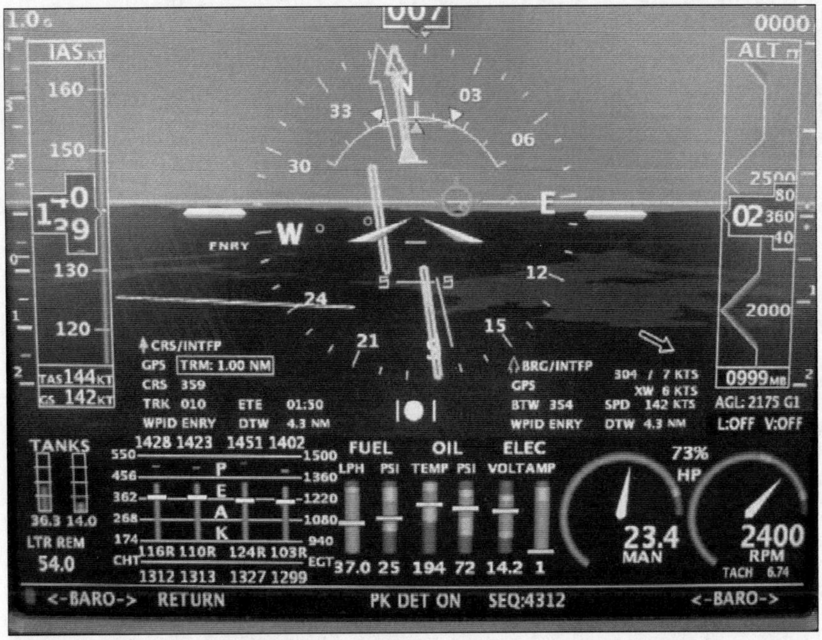

Figure 23. Advanced Flight Systems (AFS) PFD (A. O. Frog, March 2014).

AIRCRAFT CONTROLLABILITY

With reference to the previously mentioned accidents it is also worth mentioning the role of control sticks (inceptors) and handles. These should be even more intuitive than modern controls. It is important that pilots receive feedback cues from the controls which help them assessing the flight situation and giving the pilots better situational awareness without total reliance on sight and visual readings[19]. It may be argued that the PM (pilot monitoring) in AF447 could not see the control input of the PF. Hence, it is not clear if he was aware that the PF held full aft stick and was holding the aircraft in a controllable deep stall.

The pilots use several information cueing channels during aircraft control, Figure 24. The visual cues are most effective, but tactile cues are also of great importance during certain flight conditions (i. e. moving controls/inceptors give pilot feedback cues).

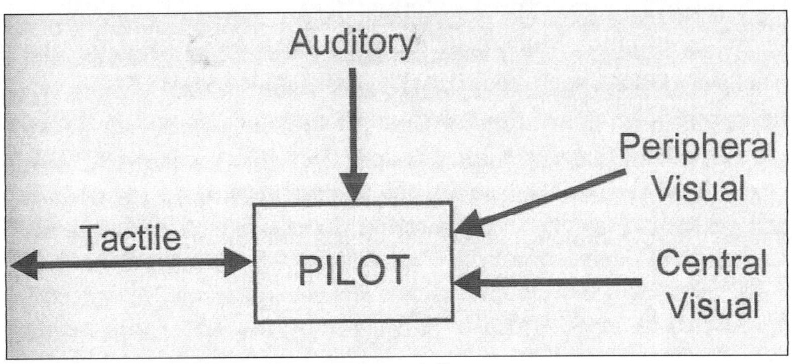

Figure 24. Pilot information cueing channels[20].

---

[19] References 9 and 10

[20] References 9

The basic piloting model is "the series pilot model" as described by Field 2004[21], Figure 25. The model describes basic piloting technique as student pilots are taught during basic training. The model is intuitive and is the most effective model during low speed manual handling of aircraft. The flight displays and controls (inceptors) should be based on this model.

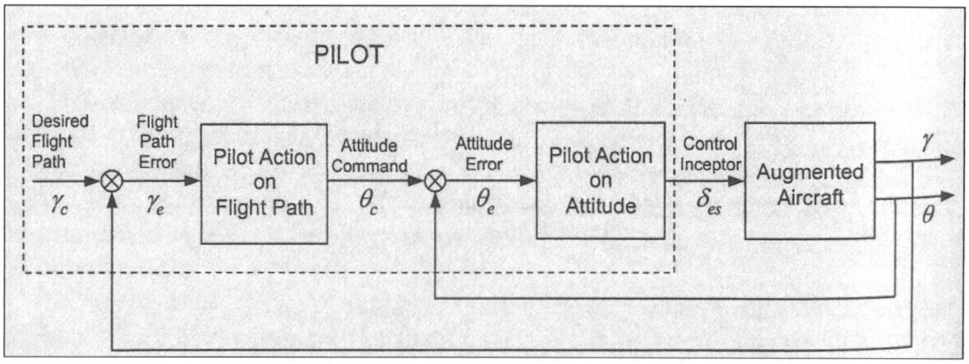

Figure 25. The series pilot model[22]

It may seem that todays pilots' knowledge about basic aircraft performance is not adequate. During the 1950-60-era, when the jet aircraft was introduced, it was focused on "operations on the back side of the power curve". This was a result of several approach accidents, both in the military and commercial aviation.

Pilots became quite familiar with the theory of aircraft performance, where pitch attitude controlled the airspeed and power controlled the rate of climb and descent:

$$\mathbf{V = \sqrt{(2L / C_L \varsigma S)}}$$

$$\mathbf{R/C = T - D\ (V/W)} \qquad \mathbf{R/D = D - T\ (V/W)}$$

Today manufacturers have made flying more complicated than necessary with too many autopilot modes and pilot's increased reliance on automation. How difficult is it to control an aircraft? Today basic aerodynamic principles seem to be forgotten by airline pilots normally controlling the aircraft through the autopilot. Modern aircraft are equipped with autopilot and autothrottles. The original intent was that pilots should use these automated systems in combination to reduce cockpit work load. When flying an autopilot coupled approach, the autopilot controlled the pitch and roll and the autothrottles controlled the rate of descent. The idea was that the PF should disconnect both when flying manual. However, some pilots developed the habit of disconnecting the autopilot and control pitch and roll manually and let the autothrottle maintain speed. Hence, the auto throttle controlled the speed and the pilot controlled the pitch and roll. Ref. Asiana Flight 214 accident.

"Pilots normally try to land at the target speed, in this case 137 knots, plus an

[21] References 10

[22] References 10

additional 5 knots, said Bob Coffman, an American Airlines captain who has flown 777s.
He said the briefing raises an important question: "Why was the plane going so slow?"
The engines were in idle and the pilots were flying under visual flight rules, Hersman said. Under visual flight procedures in the Boeing 777, the autopilot typically would have been turned off while the automatic throttle, which regulates speed, would been on until the plane had descended to 500 feet, Coffman said. At that point, pilots normally would check airspeed before switching off the autothrottle to continue a "hand fly" approach, he said."[23]

It seems that Asiana Flight 214 accident may be similar to the Turkish Airlines B737 accident in Amsterdam, where the throttles were in idle (due to intercept of glide path from above) and the autopilot was trying to compensate with increasing AOA, instead of PF handflying the aircraft manually during the final visual part of the approach. This seems to be a reversion to the type of aircraft accidents which occurred during the early 1960-ies, with the introduction of passenger jet aircraft. Older airline pilots were not familiar with jet aircraft performance characteristics with approach speeds "on the back side of the power curve".

> *"The instructor pilot told investigators that at 500 ft. altitude, he realized the aircraft was below the PAPI's visual glideslope and told the left seat pilot to "pull back" on the control yoke. "He had set the speed at 137 kt. and assumed the autothrottles were maintaining the speed," the NTSB says. Depending on the auto-flight mode selected, autothrottles, if armed and turned on, should automatically control engine thrust to maintain a preset speed, in this case 137 kt., the reference landing speed for the 777-200ER that day."*[24]

This control technique is opposite of basic piloting principle, and eventually the pilots may develop a habit of controlling the aircraft's vertical flight path in this manner. We may also see the effect of this habit in the AF447 accident, where the PF was pulling on the control stick (in stead of lowering the nose) and adding full thrust and thus trying to "fly out of the high AOA condition". This would be in line whith his previous training with Airbus systems, where he normally is protected by autothrust if approaching stall AOA (Airbus quote; *"our airplanes do not stall"*).
Hence, an AOA indicator should be a mandated supplement to the airspeed indicator, and by the same token it would be logical to have an airspeed scale with decreasing digits with increasing stick pull ("fly to - principle"), or even better, a transparent round dial as shown in Figure 12. The combination of intuitive flight displays, controls (inceptors) and autopilot modes, should be related to basic flight mechanics principles learned during basic pilot training.

DEVELOPMENTS IN PILOT EDUCATION AND TRAINING

[23] Curt Lewis & Associates, Flight Safety Information, July 15, 2013.
[24] Curt Lewis & Associates, Flight Safety Information, July 15, 2013.

Instrument flying ("*blind flying*") was first introduced by Jimmy Doolittle in USA in 1929. However, it was not until World War II that instrument flying became standard pilot proficiency. After the war instrument flying was continuously developed by improved instrumentation and pilot procedures. The concept of Attitude Flying was introduced. The basic principle is that the pilot controls the airspeed by lowering or raising the nose attitude of the aircraft. He/she controls the heading by banking in the wanted direction of flight and thus turning to the required heading. These are the basic and intuitive piloting principles both in visual and instrument flight. In visual flight the pilot uses the natural horizon as reference, and in instrument flight he/she uses the attitude indicator ("*artificial horizon*").

This concept was further developed into categories of instruments; Control Instruments consisting of Attitude and Power Instruments, Performance Instruments and Navigation Instruments. The principle was simple and intuitive; the pilot controlled the attitude of the aircraft by reference to only two instruments (Attitude and Power) and monitored the performance and navigation instruments for proper response, Figure 26.

A special pilot instrument cross check was developed, Figure 27, where the main instruments were the Attitude and Power indicators (Control Instruments). Improper cross check technique could result in chasing the performance indicators preventing stabilized flight. The student pilots were trained according to these principles and learned to scan the instruments in a certain scan pattern, where the Attitude Indicator was in the center and the most frequent scanned instrument. The faster this scan could be developed, the better situational awareness for the pilot.

During basic and advanced flight training, both in visual and instrument flight, it was emphasized that the primary Flight Control Instruments were the Attitude Indicator and the RPM/Power instrument – Pitch controlles airspeed and Power controlles acceleration, climb and descent according to:

$$\mathbf{V = \sqrt{(2L / C_L \varsigma S)}}$$

$$\mathbf{R/C = T - D\ (V/W) \qquad R/D = D - T\ (V/W)}$$

This knowledge was also the fundamental basis for controlling an aircraft during any emergency or upset flight condition:

- Maintain aircraft control (by use of attitude and power) (Aviate)
- Analyze the situation (Navigate)
- Take proper action (Communicate)

Today's pilots seem to forget this, possibly because authorities and manufacturers allow design and operational use of conflicting control laws and operational practice.

Figure 28 shows the T-shape instrument scan of modern strip PFD. This scan will take longer as the eyes must be directed to one focal point at a time and the pilot fails to see more than one parameter at a glance as opposed to round dial instruments and pointers. With larger round dials and pointers the pilot may use his peripheral vision giving him/her an overall quicker scan.

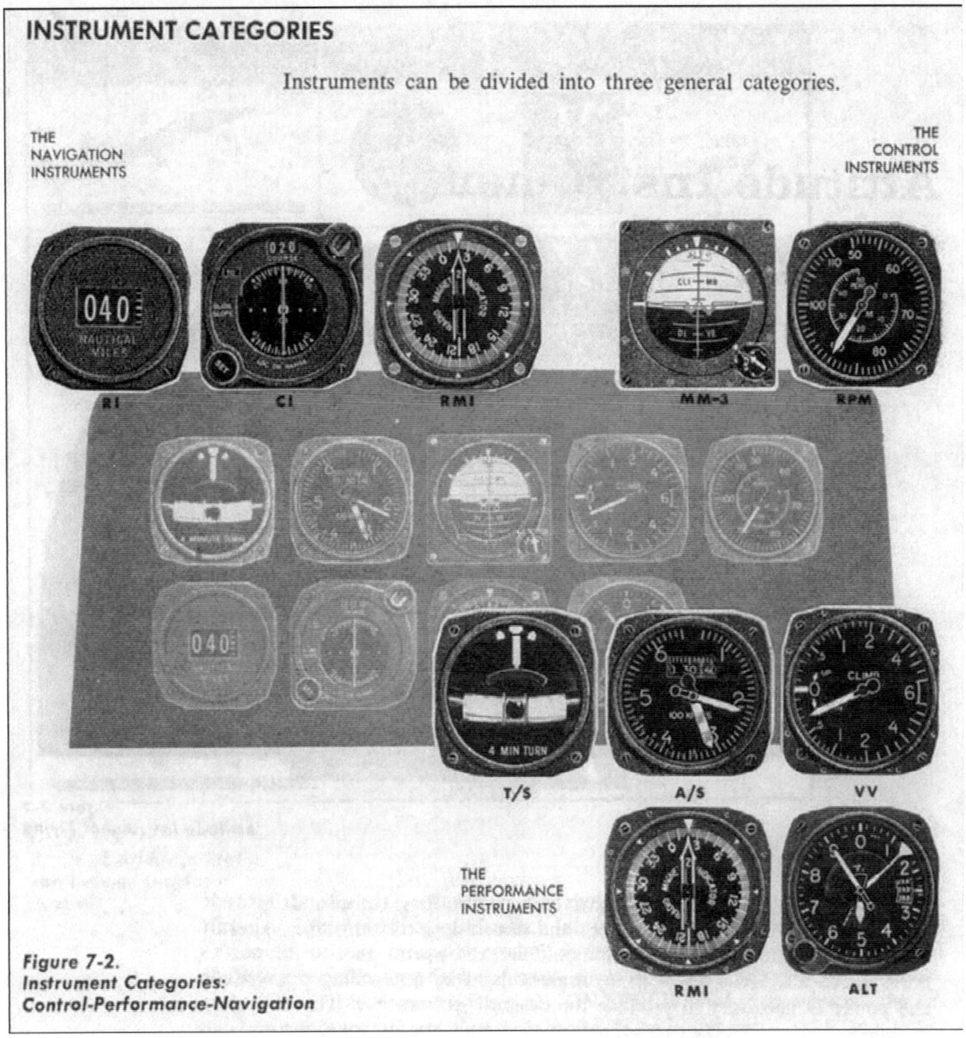

Figure 26. Flight instrument categories (US Air Force Manual 51-38, 1954).

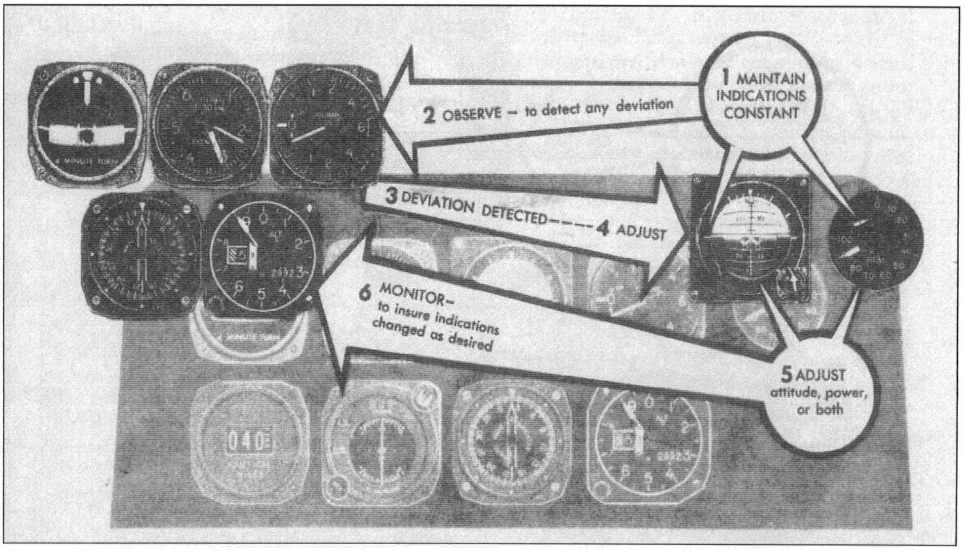

Figure 27. Instrument cross check technique (US Air Force Manual 51-37, 1979).

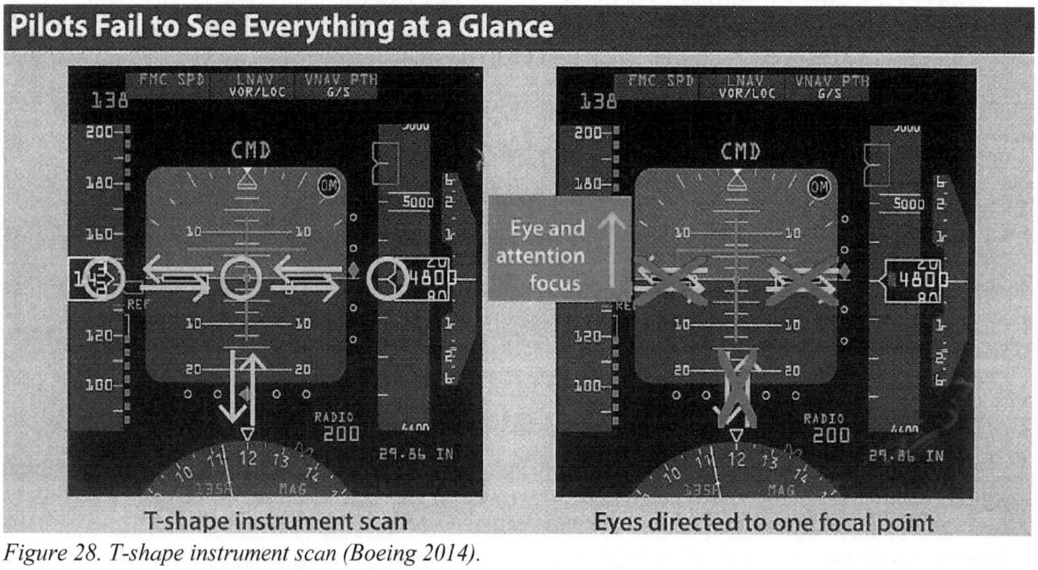

Figure 28. T-shape instrument scan (Boeing 2014).

COCKPIT DISPLAYS – TEST AND EVALUATION

Reference 20 addresses test and evaluation of cockpit displays:

> *"There have been a number of papers and articles written about operational difficulties with modern display and other cockpit systems. As we see it, the problem has been a series of discontinuities between the users and the designers, between the designers and the testers, and between the users and the testers.*
> *As a result of the first discontinuity, between users and designers, inadequate design requirements are established. This is particularly unfortunate as systems can be (and are being) designed with greater and greater capabilities in terms of automatic flight and guidance and flight control. Without adequate requirements, it is hardly surprising that there are problems encountered in operational use.*

The second discontinuity, between designers and testers, reduces the opportunities for feedback to the designer. In fact, with today's economic setting, many systems are practically committed to production by the time they reach flight test. Only if there are very serious problems, will these systems be corrected.

The third discontinuity, between users and testers, results in inadequate test criteria. As a result of this discontinuity, we are left with highly subjective criteria which vary from tester to tester. Or we have inappropriate criteria."

*"**The display design** must consider **why the pilot needs the data** and **what the pilot is expected to do with the data**. According to Singleton, several questions must be answered during development of a display:*

- *Does the **pilot's need** justify the display?*
- ***What data does the pilot need** that has not been provided?*
- *Can the **average pilot** obtain what is required easily?*
- *Does the **display conform to the real world**?*
  - *To other cockpit displays?*
  - *With previous pilot habits and skills?*
  - *With required decisions and actions?"*

RESEARCH ON PFD AND INCLUSION OF AOA INDICATOR

Figure 29 is described in Reference 4 as the results of a research program related to readability of different types of airspeed and altitude displays on HUD's.

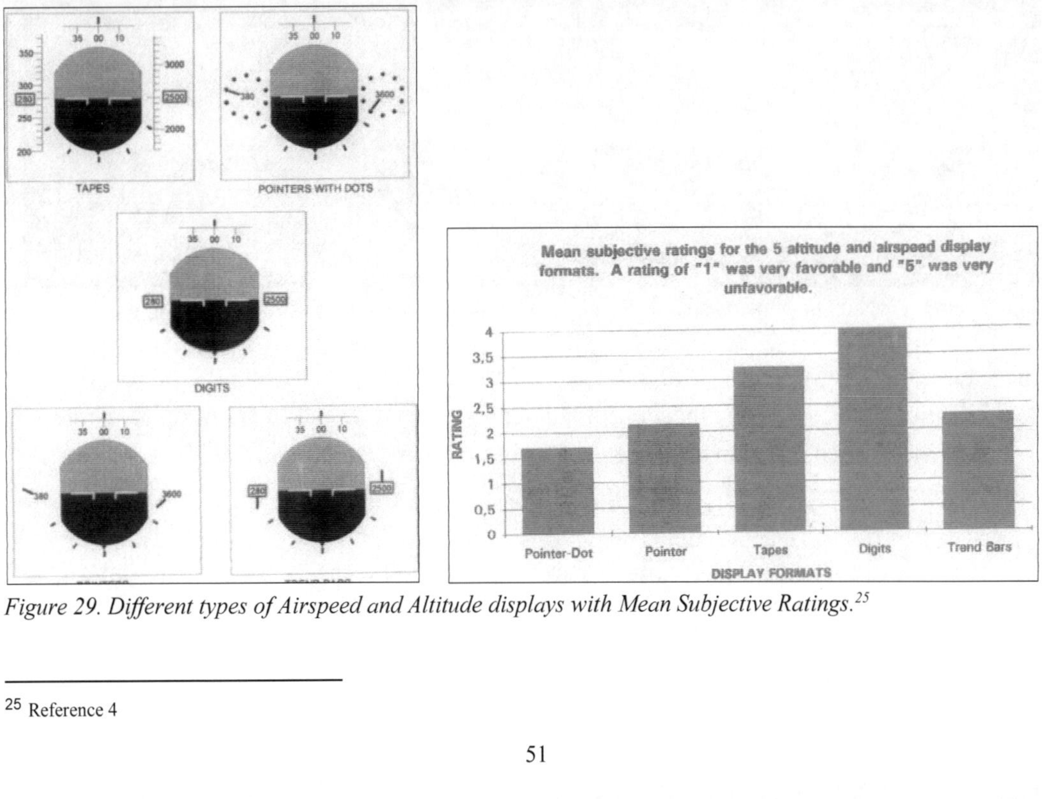

Figure 29. Different types of Airspeed and Altitude displays with Mean Subjective Ratings.[25]

[25] Reference 4

We see that a pointer is more intuitive and quicker to read and interpret at a glance. This indicates that round dial and pointer is more human adaptive than tapes and digits.

Figure 30 shows a PhD study at the Swinburne's Aviation Simulation Laboratory, Swinburne University of Technology's, Australia. The initial results into a study of the way pilots look at digital and analog instruments suggest experienced pilots facing an emergency will spend twice as much time looking at their instruments as novices. The study is testing volunteers in Swinburne University of Technology's flight simulators to see how pilots cope with the switch between digital and analog cockpits.

There is evidence indicating that pilots who have trained on analog instruments generally find it easier to move to a digital cockpit than vice versa. The author has not seen any research data supporting the human factor benefits of the state-of-the-art PFD's of today with strip indicators. However, there are indications that digital round dial indicators are more intuitive and quicker to read at a glance than strip indicators. The PFD's of today are 30 years old and it is time to develop new PFD based on human factor research and technological advancements.

Figure 30. PhD study at Swinburne's Aviation Simulation Laboratory, Australia 2014. Source: Supplied.

In the future there might be novel aircraft control systems available to help the pilots in a LOC-I situation [26] but so far the most rational solution is to make the PFD, autopilot modes and control inceptors more human factor centric and intuitive, combined with a prominent AOA indicator and relevant pilot training in stall and upset recovery.

[26] Reference 24

CONCLUSION

Several recent LOC-I accidents indicate that the human factor related aspects of primary flight displays, lack of AOA displays, flight control and inceptor design, and pilot education and training[27], play a significant role in the pilot's handling of an aircraft and loss of situational awareness.

It is recommended that test pilots and SETP members get more involved in human factor based research and development of aircraft controllability, autopilot modes, AOA and more intuitive primary flight displays.

Quote: *"If you keep on doing what you do – you get what you got".*[28]

[27] Curt Lewis & Associates, Flight Safety Information, October 16, 2014

[28] Captain Tore Hultgren, AIBN

ABBREVIATIONS

AAIB - Aircraft Accident Investigation Branch (UK)
ACAS - Airborne Collision Avoidance System
ADI - Attitude Director Indicator
AI - Attitude Indicator
ALA - Approach and Landing Accidents
AOA - Angle of Attack
ASAR – Arc Segmented Attitude Reference
GA – General Aviation
CFIT - Controlled Flight into Terrain
CS-25 - Certification Specification Part 25 (EASA)
DCIEM - Defence and Civil Institute of Environmental Medicine Canada
EASA- European Aviation Safety Authority
EFIS – Electronic Fligh Instrument System
EGPWS - Enhanced Ground Proximity Warning System
FAA - Federal Aviation Administration
FAR 25 - Federal Air Regulations Part 25 (US)
GPWS - Ground Proximity Warning System
HMI – Human Machine Interface
HSI - Horizontal Situation Indicator
HUD - Head Up Display
IAS – Indicated airspeed
IATA - International Air Transport Association
ICAO - International Civil Aviation Organization
IFIS - Integrated Flight Instrument System (USAF)
ITQI - IATA's training and qualification initiative
LOC-I - Loss of Control in the air type of accidents
MMO - Max Operating Mach number
ND - Navigation Displays
NGAP - ICAO's next generation aviation professionals
NTSB - National Transport Safety Board
PF - Pilot Flying
PFD - Primary Flight Displays
PM - Pilot Monitoring
PVD - Peripheral Vision Display
TAWS - Terrain Avoidance and Warning System
TCAS - Traffic and Collision Avoidance Systems
UA - Unusual attitude
UR – Upset revovery
VD - Vertical Displays
VMO - Max Operating airspeed
VRS - Vortex Ring State
VSI - Vertical Speed Indicator
WR - Weather Radars

REFERENCES

[1] USAF Test Pilot School, *Stability and Control Handbook*, FTC-TIH-68-1002 1968.

[2] NASA Technical Memorandum 74097, *Aerodynamic Characteristics of Airplanes at High Angles of Attack.* Joseph R. Chambers and Sue B. Grafton, NASA Langley Research Centre 1977.

[3] Svimonoff, C. *The Air Force Integrated Flight Instrument Panel.* WADC TR 58-431, October 1958.

[4] Ercoline, W and Gillingham, K. *Effect of Variations in Head-Up Display Airspeed and Altitude Representations on Orientation Recognition.* Proceedings of the Human Factors Society 34th Annual Meeting 1990.

[5] Reising, M. J., Butterbaugh, L. C., Liggett, K. K. *Preliminary Assessment of primary Flight Display Symbology for Electro-optic Head-Down Displays*, June 1991.

[6] *Human Factors Team Report on the Interfaces between Flight Crews and Modern Flight Deck Systems.* FAA 1996.

[7] Lovesay, E. J. *The Instrument Explosion – a Study of Aircraft Cockpit Instruments.* Applied Ergonomics 1997.

[8] Lande, K. *Standardization of Flight Decks - Operational Aspects.* Soekkha, H (Ed). Aviation Safety - Human Factors - System Engineering - Flight Operations - Economics - Strategies - Management, pp. 189-201. Proceedings of the International Aviation Safety Conference 1997 (IASC-97). VSP BV 1997.

[9] Harris, D. *Head-down flight deck display design.* Harris, D (Ed). Human Factors for Civil Flight Deck Design, pp. 69-102. Ashgate 2004.

[10] Field, E. *Handling qualities and their implications for flight deck design.* Harris, D (Ed). Human Factors for Civil Flight Deck Design. pp. 157-181. Ashgate 2004.

[11] Brown, A. P., Dillon, G. C. J., Erdos, R. *Flight Manoeuvres and Spin Characteristics of the Harvard Mk 4: Application to Human Factors Flight Research.* AIAA 2004-4815, 2004.

[12] Leslie, T. *NRC Canada Harvard Unusual Attitude Recovery Research Project.* AB-FRL-2002-0045, June 2002.

[13] AC 25-11A. *Electronic Flight Deck Displays.* FAA 2010.

[14] Dekker, S. *The Field Guide to Understanding Human Error.* Ashgate 2006.

[15] Reason, J. *Managing the Risks of Organizational Accidents.* Ashgate 1997.

[16] Bennett, S. Dr. *The Air France 447 Accident – Organizationally and Socially Produced?* RAeS, The Aerospace Professional, p. 20-21. October 2012.

[17] Lande, K. *Causes for Pilot Mistakes - Aircraft Controllability - Cockpit Design - Flight*

Displays - Basic Pilot Training. SETP Flight Test Workshop, Salzburg, November 2012.

[18] Graham H. Richard, Col. USAF (Ret). *Flying the SR-71 Blackbird*, Zenith Press 2008.

[19] Coombs L. F. E., *The Aircraft Cockpit, from stick-and-string to fly-by-wire,* Patrick Stephens 1990.

[20] Newman Rickhard L. and Greeley Kevin W., *Cockpit Displays: Test and Evaluation,* Ashgate 2001.

[21] Bateman Don. *Some Thoughts on Reducing the Risk of Aircraft Loss of Control.* Advanced Technology, Honeywell International, Redmond WA. Paper not dated.

[22] Introduction of Glass Cockpit Avionics into Light Aircraft. Safety Study NTSB/SS-01/10 PB2010-917001, 2010.

[23] Lande, K. *Aircraft Controllability and Primary Flight Displays*. SETP Symposium, Luleå, June 2014.

[24] Stoop John, de Kroes Jan. *Loss of Control: An Inherent Frontier for Managing Recilience?* Kindunos Safety Consultancy Ltd. The Netherlands 2015.

[25] Lande, K. *Aircraft Controllability and Primary Flight Displays - A Human Factor Centered Approach.* SETP USA Central Section Regional Symposium, Wichita, KS, 18-19 June 2015.

[26] Lande, K. *Black Swans or Causes for Pilot Mistakes*. EFTSW Aix-en-Provence, France 3-5 November 2015.

[27] Cheung, Bob DR. *Non-visual Spatial Orientation Mechanisms.* Defence and Civil Institute of Environmental Medicine (DCIEM) Canada, not dated.

[28] Mumaw Randall J. Dr. et.al. *New Tool for Analysing the Potential Influence of Vestibular Illusions.* January-March 2016 ISASI Forum.

AND THE WORLD TURNED: SPIN TESTING THE DG-1000S

Tim McDonald (AF)
USAF Test Pilot School

Presented to
The Society of Experimental Test Pilots
January 2016

Distribution: A. Approved for public release; distribution unlimited. 412 TW-PA-15656

Figure 1. Spin Testing over Roger's Dry Lake, Edwards AFB, CA.

INTRODUCTION: RESPECTING THE DISCIPLINE

During the past 10 years, the Society of Experimental Test Pilots (SETP) has published over 250 new technical papers in its online database. Of these, only four of them – less than 2% -- directly address the topic of spin testing. Since it apparently is not a test discipline many of us encounter, it might seem logical to question the importance of remaining knowledgeable of its precepts. If you did question its importance, recent experience suggests you would not be alone: experienced SETP members have even gone so far as to openly question whether or not the USAF Test Pilot School (TPS) should continue training its students on the topic.

A survey of the lessons-learned from the SETP spin papers that *are* available – when compared to the lessons-learned of this project – reveals a striking recurrence of themes. Seemingly minor aberrations in the aircraft mold line, for example – particularly on or near the nose of the aircraft or wing leading edge – have led to the discovery of unpredicted out-of-control behaviors and potentially-dangerous recoverability characteristics.[1] In the face of such uncertainties, selecting an effective build-up approach has proven problematic in some cases.[2]

Even when not engaged in stall, departure susceptibility and/or spin testing, test pilots routinely perform maneuvers that have led to sudden and unpredicted departures from controlled flight. Stability and control testing with large lateral asymmetries opens the door to such a possibility, for example.[3] SETP members have recently listened to presentations describing unexpected departures during seemingly benign maneuvers: the unintended inverted spin experienced by the *SpaceShipTwo* test crew – immediately following release from *WhiteKnightTwo* – provides a compelling example.[4]

One of the goals of this paper, then, is to refresh understandings that could be needed on a moment's notice. It examines spin behaviors which were previously undocumented and unusual, identifies unexpected hazards, offers techniques to mitigate them, and proposes a new method of assigning spin mode modifiers for highly/violently oscillatory spins. It also informs the reader of what recent USAF TPS graduates know – and by inference, don't know – about this field of testing, where such insight is prerequisite to effective risk mitigation. Finally, it is hoped it will serve as a readily-available primer on the topic of spin testing, while helping to banish certain preconceived ideas and attitudes which consistently prove self-critiquing in the flight test profession. On occasion, they even prove fatal.

UNDERSTANDING THE USER REQUIREMENT

As part of its mission to graduate full-spectrum test professionals, TPS has used the ASK-21 sailplane to conduct spin test training for over 20 years: over 1000 testers have been trained in this aircraft to date. First spin-tested for the USAF by Charles Precourt (F) in 1989[5], the ASK that is available to TPS is approaching the end of its useful lifespan. The TPS Flying Qualities Branch required a replacement aircraft that would facilitate an overall learning objective to "practice making accurate observations in a highly dynamic flight environment" – as well as several specific learning objectives. These included: 1) Using Phase A-D stalls to determine the aircraft's MIL-F-83691B susceptibility to departure, 2) Identifying MIL-F-83691B spin phase boundaries, 3) Identifying MIL-F-83691B Spin Mode Modifiers, 4) Investigating the efficacy of various spin recovery procedures and 5) Investigating control effects while in the spin. The Flying Qualities Branch also wanted to investigate the feasibility of further-developing students' observational skills by exposing them to inverted spin testing.

The Precourt spin test was performed in the wake of a fatal ASK-21 stall/spin mishap at the U.S. Air Force Academy. Testing revealed numerous characteristics that were not described in the published flight manual. Since that time, the USAF has elected to independently spin-test certified sailplanes prior to using them in their training curricula. In light of TPS' intentions to employ test training maneuvers that would not necessarily have been performed during initial certification testing, such an evaluation seemed especially pertinent.

While MIL-F-83691B was formally cancelled "without replacement"[6] in 1996, no substantive means of characterizing departure susceptibility or spin mode behavior has risen to take its place. TPS has elected to continue teaching its conventions and test procedures. Per its language, a departure "may be characterized by a divergent, large amplitude, un-commanded aircraft motion." What constitutes "large amplitude" appears to be left to sound engineering judgement: for test training, the test team declared a departure event at the moment any of the following parameters were exceeded: 1) 30 degrees of pitch 2) 60 degrees of roll, or 3) 90 degrees of heading change. It is important to note that measurement of these post-stall excursions began immediately after any aggravating control input was neutralized. The test team also defined "divergent" to mean that pitch, roll or yaw angles were continuing to change at a steady or increasing rate when these criteria were exceeded. See Table 1 for adapted MIL-F-83691B Phase A-D Stall descriptions and departure susceptibility ratings. Maneuvers were adapted to accommodate the absence of an Angle of Attack (AoA) display in the DG-1000S cockpit: indicated airspeed bleed rates – not AoA rates – were used. To emulate the MILSTD's defined "smooth" and "abrupt" AoA-rate test procedures, 1 kt/sec and 5 kt/sec airspeed bleed rates were used, respectively. The 5 kt/sec entry was

Stalls	Maneuver Description and Departure Susceptibility Rating
Phase A	<u>**Hold specified airspeed bleed rate until:**</u> **a)** Definite g-break, or **b)** a rapid, un-commanded angular motion, or **c)** the aft stick stop has been reached and no other indication of stall has occurred, or **d)** sustained intolerable buffet. If a departure occurs during a Phase A stall, the aircraft is rated **"extremely susceptible"** to departure.
Phase B/C	Phase A stall with one/three second(s) aggravating rudder, respectively. If a departure *first* occurs during a Phase B stall, the aircraft is rated **"susceptible"** to departure. If a departure *first* occurs during a Phase C stall, the aircraft is rated **"resistant"** to departure. If the aircraft does not depart during any Phase A-C stall, it is rated **"extremely resistant"** to departure.
Phase D	Deliberate spin attempt.

Table 1. MIL-F-83691B Phase A-D Stall Definitions

Spin Phase	Definition
"Incipient"	The initial, transitory phase of the motion during which it is not possible to identify the spin mode.
"Developed"	The phase of the spin during which it is possible to identify the spin mode.
"Fully Developed"	Attained when the trajectory has become vertical and no significant change is noted in the spin characteristics from turn to turn.

Table 2. MIL-F-83691B Spin Phase Definitions

<table>
<tr><th>Sense</th><th>Oscillations</th><th colspan="2">Attitude</th><th colspan="2">Rate</th></tr>
<tr><td rowspan="4">The sign of AOA (derived from pitch attitude).

Positive or negative AOA indicates an upright or inverted spin, respectively.</td><td rowspan="4">The qualitative magnitude of three-axis oscillations, rated by pilot comments as:
- "Smooth"
- "Mildly Oscillatory"
- "Oscillatory"
- "Highly oscillatory"
- "Violently Oscillatory"</td><td colspan="2">The magnitude of the average AOA (α_{avg}), further rated as follows:</td><td colspan="2">The body-axis yaw rate (R), further rated as follows:</td></tr>
<tr><td>"Extremely Steep"</td><td>Stall $\alpha_{avg} < \alpha_{avg} < 35$ deg</td><td>"Slow"</td><td>$R \leq 60$ deg/sec</td></tr>
<tr><td>"Steep"</td><td>35 deg $\leq \alpha_{avg} < 70$ deg</td><td>"Fast"</td><td>60 deg/sec $< R \leq 120$ deg/sec</td></tr>
<tr><td>"Flat"</td><td>$\alpha_{avg} \geq 70$ deg</td><td>"Extremely Rapid"</td><td>$R \geq 120$ deg/sec</td></tr>
</table>

Table 3. MIL-F-83691B Spin Mode Modifiers

also selected to allow us to evaluate the impact of student error when attempting to set a 1 kt/sec bleed rate, as well as verify that the 1 kt/sec entry produced the most repeatable test results. Table 2 provides a review of MIL-F-83691B spin phase definitions: it is important to note the "turn to turn" qualifier for the fully-developed phase. Table 3 provides a review of MIL-F-83691B ratings for spin mode modifiers: again, with no AoA display available in most sailplane cockpits, TPS students have been trained to visually estimate α_{avg} using pitch attitude. Under this scheme, < 20 degrees of nose-down pitch is rated as "flat," where > 55 degrees of nose-down pitch is rated as "extremely steep." Intermediate pitch angles are rated "steep."

TRANSLATING REQUIREMENTS INTO A TEST

See Figures 2 and 3. The Glaser-Dirks DG-1000S presented itself as a promising candidate replacement aircraft for the ASK-21, largely due to its unique tail and cockpit ballasting system. This system permits some degree of control over aircraft center-of-gravity (CG) and pitch moment of inertia (I_{yy}), independent of aircrew weight. This suggested the potential for repeatable spin behaviors and recovery characteristics during test training.

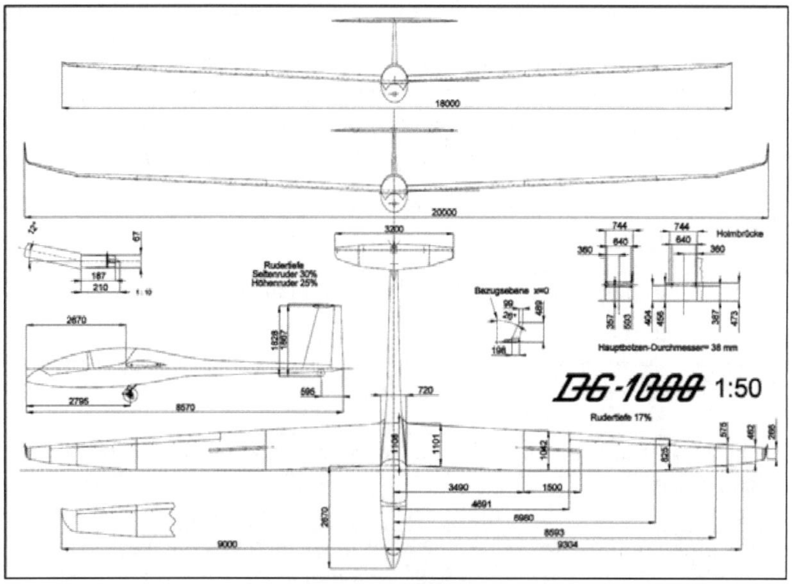

Figure 2. DG-1000 Dimensions

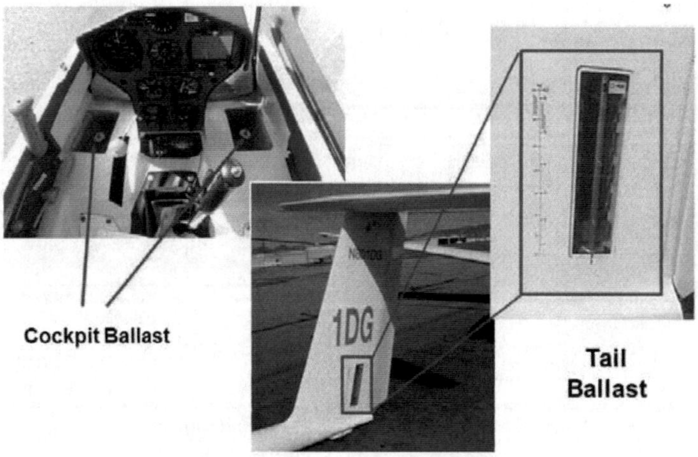

Figure 3. DG-1000S Ballast Compartments

The DG-1000S is a two-place, tandem, high performance sailplane that can be configured with an 18-meter or 20-meter wingspan. Best L/D in the 20-m configuration is 46:1. It features a mid-mounted wing, T-tail, conventional reversible flight controls, and spoilers. The undercarriage consists of a high spring-mounted electrically-retractable main wheel with disc brake and a tail wheel. The wing is constructed from carbon-fiber reinforced plastic, while the fuselage was mainly glass-fiber reinforced plastic, with steel as necessary for the landing gear. Load factor limits are +7/-5g. Manufactured in Germany, the DG-1000S is certified under the airworthiness requirements of the *Joint Aviation Authorities,* (JAR) *Joint Requirements JAR-22 Sailplanes and Power Sailplanes* for spins in the Utility category for both wing configurations; however, inverted spins are certified only in the 18m configuration.

TPS learning objectives translated readily into specific test objectives: the test team aimed to evaluate how well – and safely – the aircraft illustrated those characteristics TPS students needed to observe and assess. The test team also anticipated the need to evaluate the impact of student control input errors on upright spin recovery characteristics – henceforth referred to as a "misapplied controls" investigation. This investigation was not deemed necessary for inverted spin testing, as inverted spins were anticipated to be an instructor demonstration during a curriculum event. Training requirements did not mandate an evaluation of accelerated stall/spin entries.

Once test objectives had been appropriately narrowed to meet these requirements, identifying data requirements for the test was relatively straightforward. Since the focus of the training event was developing student observational skills in a spin, test pilot observations constituted the primary data source. To remain legally certified, no permanent modifications to the (privately owned) aircraft were permitted: this prohibited us from installing the necessary instrumentation to precisely measure parameters like AoA, Angle of Sideslip, airspeed, or altitude. Moreover, defined test training requirements did not explicitly require this data. See Figure 4. Lacking a cockpit ADI, markers were placed on the canopy to help visually identify 20 degrees and 55 degrees of nose-down pitch – to facilitate identification of the "attitude" spin mode modifier. A yaw string provided the pilot an approximation of sideslip as well as AoA trend information.

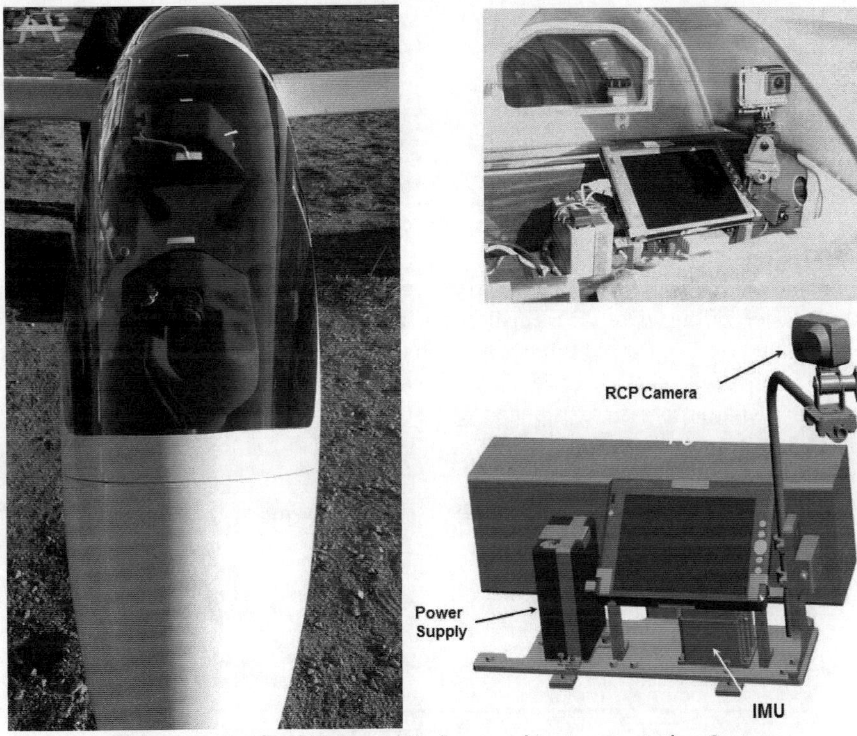

Figure 4. DG-1000S Canopy Markers and Instrumentation System

Thanks to the proliferation microelectromechanical systems (MEMS) technology, the test instrumentation team was able to supplement pilot observations with a compact, lightweight instrumentation system affixed to a bulkhead immediately behind the rear cockpit. The pallet included a MEMSIC NAV-440 inertial measurement unit (IMU) which was capable of measuring body-axis pitch, roll and yaw angles, derived body-axis angular-rates, inertial velocity and load factor. An internal attitude-heading-reference-system (AHRS) also provided aircraft heading and heading-rate information using a 3-axis internal magnetometer.[7] IMU control and data recording was handled via tablet PC. Front and rear digital cameras also captured cockpit instrument readings and out-the-window heading and attitude references for post-flight review.

REVIEWING THE THEORY; SPEAKING THE LANGUAGE

See Figure 5. In a spin, both wings are generally beyond stall angle of attack, but one wing is at a higher AoA than the other. This more completely-stalled wing experiences less lift and greater induced drag – causing it to "retreat." The other wing simultaneously experiences more lift and less drag, causing it to "advance." This is the aerodynamic forcing function for autorotation.

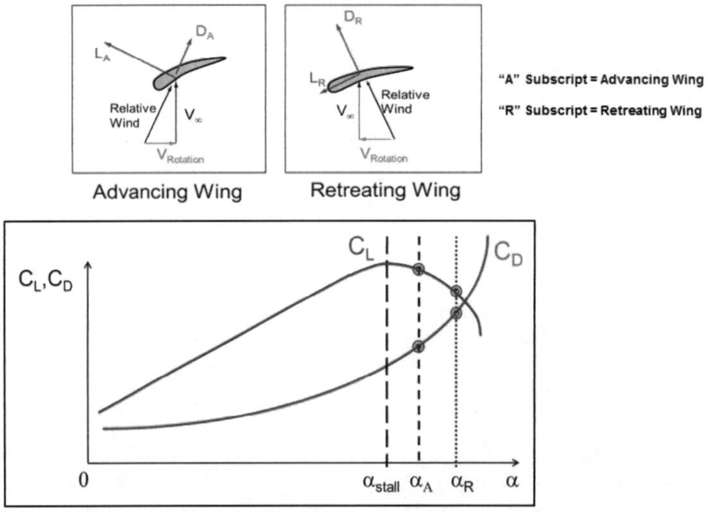

Figure 5. Spin Aerodynamics

See Figure 6. An aircraft with positive static stability experiences an aerodynamic nose-down pitching moment in a spin, as depicted on the left of the figure. Mass distribution along the length of the fuselage – coupled with centripetal acceleration (depicted on the right) tends to pitch the nose back up. Shortly after a departure, these forces are often not in balance with each other, leading to incipient-phase oscillations in pitch, roll and yaw. In a fully-developed spin, aerodynamic forces are balanced against equally-opposing inertial forces, producing essentially constant spin parameters "turn to turn."

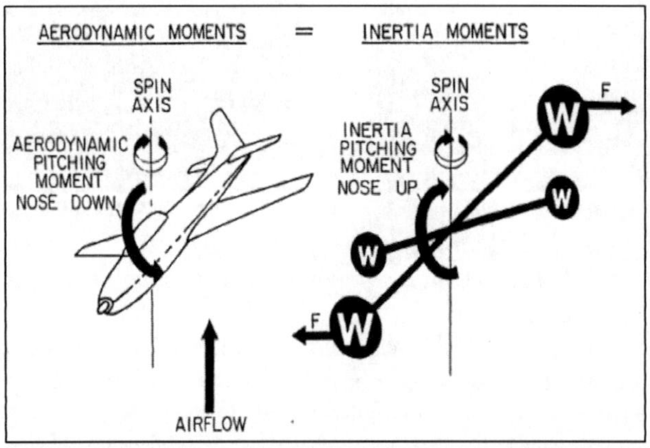

Figure 6. The Balance of Aerodynamic and Inertial Forces in a Spin

The interaction between aerodynamic and inertial forces can also be understood via review of aircraft equations of motion. X_b, Y_b, and Z_b refer to the orthogonal body axis frame of reference, centered on the aircraft CG. See Figure 7: specific orientation of this axis system will be addressed shortly. *P, Q,* and *R* refer to body-axis roll, pitch and yaw rates, respectively. *L, M* and *N* refer to body-axis roll, pitch and yaw moments, respectively. $\vec{H}$ refers to total angular momentum, where $\vec{M}$ refers to total moment – both about the CG. $\vec{I}$ represents inertia about a specific body axis (I_{xx}, I_{yy}, I_{zz}) or a product of inertia between two specific body-axes (I_{xz}, for example). Recall the moments of inertia describe an aircraft's resistance to rotation about its respective axis, where products of inertia describe how a moment applied to one axis changes angular momentum in another. Products of inertia, then, provide a measure of the asymmetry of mass distribution on either side of a body-axis plane. $\vec{\omega}$ represents angular velocity.

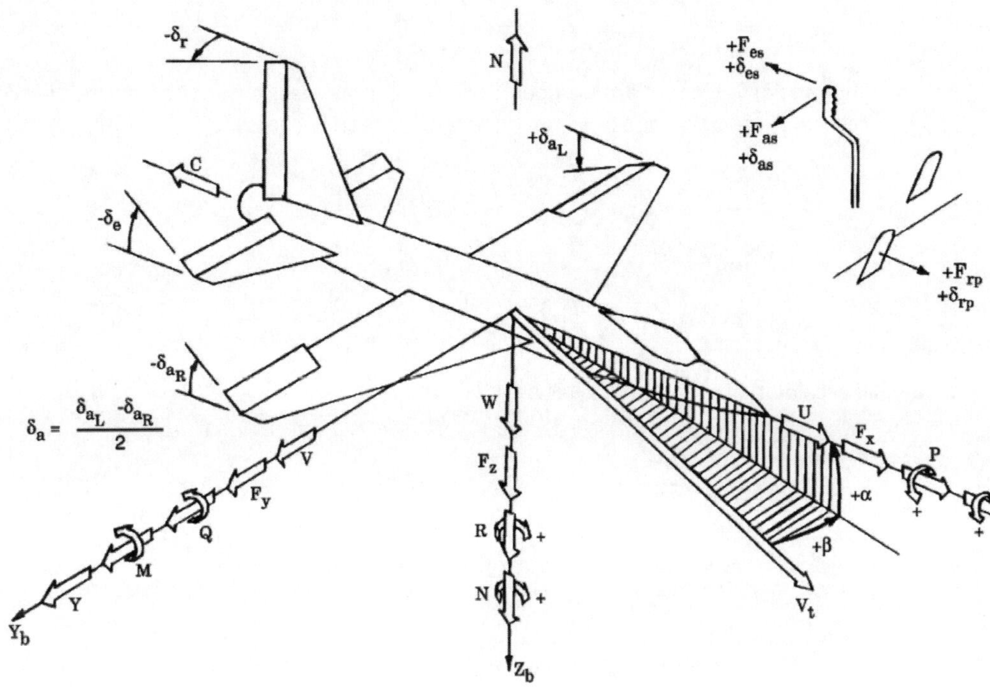

Figure 7. Body-Axis Definitions and Sign Conventions

Derived equations of motion for a spin rest upon several important assumptions:

1. *An inertial reference frame is used.*
2. *The aircraft can be treated as a rigid body in translation and rotation.*
3. *The mass of the aircraft remains constant during the maneuver.*
4. *A flat, non-rotating Earth is a suitable inertial reference frame.*
5. *Aircraft motion will be described using a body-fixed reference frame.*
6. *The atmosphere is at rest relative to the inertial reference frame.*
7. *Thrust effects are negligible for a sailplane.*
8. *The aircraft has symmetry about the XZ plane. Therefore,* $I_{yz} = I_{xy} = 0$.

From Newton's Second Law, the sum of applied forces equals the time-rate-of-change of linear momentum. The rotational equivalent of this concept for spins is similar: the sum of applied aerodynamic moments equals the time-rate-of-change of angular momentum, or:

$$\left[\frac{d\vec{H}}{dt}\right]_{Inertial} = \Sigma(\vec{M}_{aero})$$

Where,

$$\vec{H} = \vec{I}\vec{\omega}$$

Therefore,

$$H_{Inertial} = \begin{bmatrix} I_{xx} & -I_{xy} & -I_{xz} \\ -I_{yx} & I_{yy} & -I_{yz} \\ -I_{zx} & -I_{zy} & I_{zz} \end{bmatrix} \begin{bmatrix} P \\ Q \\ R \end{bmatrix}$$

With I_{yz} and I_{xy} = 0, (assumption #8) four terms are dropped from the inertia tensor above. The balance between aerodynamic moments and time-rate-of-change of angular momentum then becomes:

$$\dot{P}I_{xx} + QR(I_{zz} - I_{yy}) - (\dot{R} + PQ)I_{xz} = L_{aero}$$
$$\dot{Q}I_{yy} - PR(I_{zz} - I_{xx}) + (P^2 - R^2)I_{xz} = M_{aero}$$
$$\dot{R}I_{zz} + PQ(I_{yy} - I_{xx}) + (QR - \dot{P})I_{zx} = N_{aero}$$

Solving these for angular acceleration in each axis gives us:

$$\dot{P} = \frac{I_{yy} - I_{zz}}{I_{xx}} QR + \frac{I_{xz}}{I_{xx}}\left(\dot{R} + PQ\right) + \frac{L_{aero}}{I_{xx}}$$

$$\dot{Q} = \frac{I_{zz} - I_{xx}}{I_{yy}} PR - \frac{I_{xz}}{I_{yy}}\left(P^2 - R^2\right) + \frac{M_{aero}}{I_{yy}}$$

$$\dot{R} = \frac{I_{xx} - I_{yy}}{I_{zz}} PQ + \frac{I_{xz}}{I_{zz}}\left(\dot{P} - QR\right) + \frac{N_{aero}}{I_{zz}}$$

We will now orient the body axis so that it aligns with the inertial axis – by assuming an X-Y plane of symmetry. This leads to our final assumption:

9. *The aircraft body axis system is chosen to coincide with the aircraft inertial axis. Therefore, I_{xz}=0.*

In most sailplanes, the vertical tail accounts for the most obvious – and relatively minor – asymmetry across the X-Y plane. This final assumption will cancel all products of inertia. While approximate, this is close enough to the truth to illuminate important characteristics. The equations now reduce to:

$$\dot{P} = \frac{I_{yy} - I_{zz}}{I_{xx}} QR + \frac{L_{aero}}{I_{xx}}$$

$$\dot{Q} = \frac{I_{zz} - I_{xx}}{I_{yy}} PR + \frac{M_{aero}}{I_{yy}}$$

$$\dot{R} = \frac{I_{xx} - I_{yy}}{I_{zz}} PQ + \frac{N_{aero}}{I_{zz}}$$

For a wing-loaded aircraft like the DG-1000S, $I_{xx} > I_{yy}$. Also recall that I_{zz} is the largest inertia term. Therefore, the sign of the inertial terms are as shown:

$$\dot{P} = \overset{(-)}{\frac{I_{yy} - I_{zz}}{I_{xx}}} QR + \frac{L_{aero}}{I_{xx}}$$

$$\dot{Q} = \overset{(+)}{\frac{I_{zz} - I_{xx}}{I_{yy}}} PR + \frac{M_{aero}}{I_{yy}}$$

$$\dot{R} = \overset{(+)}{\frac{I_{xx} - I_{yy}}{I_{zz}}} PQ + \frac{N_{aero}}{I_{zz}}$$

From this final result, note the effect of pitch rate (Q), yaw rate (R), and roll rate (P) on the aircraft response. For example, the product of P and R couple to produce pitch acceleration. This is commonly referred to as "inertial coupling." Loadings with greater pitch inertias (I_{yy}) generate greater inertial pitch up moments, allowing spins to stabilize at a more forward C.G. despite the greater aerodynamic pitch down moments inherent to that C.G. location. The DG-1000S ballasting system provides direct control over this inertia term, thereby permitting effective spin test training across the broadest-possible range of aircrew weights.

Finally, recall that one of the TPS learning objectives was to allow students to investigate control effects while in a spin. Given the wing-loaded nature of a sailplane, an aileron effects investigation provides an effective way to accomplish this. See Figure 8. In this tail-view of a sailplane in a left spin, imagine the mass distribution along the lateral axis to be something like a dumb bell, with each depicted mass following a circular path around the spin axis of rotation. With ailerons retaining some effectiveness in a spin, the aircraft will bank left when left aileron is applied, effectively pulling the two masses toward the center of rotation. Conservation of angular momentum dictates an immediate increase in the rate of rotation. From the pilot station, this appears as in increase in the time-rate-of-change of heading – often reported by the pilot as in increase in yaw rate. In the body axis frame of reference, however, the lateral input manifests itself as in increase in body-axis yaw and pitch rate. Paying careful attention to the sign conventions found in Figure 7, our finalized equations for $\dot{Q}$ and $\dot{R}$ confirm this.

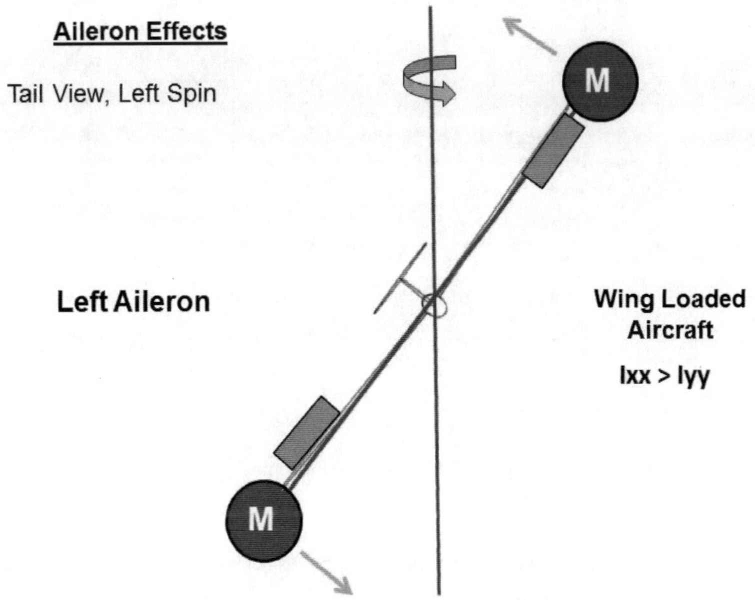

Figure 8. Aileron "Into" a Spin Increases Body-Axis Yaw & Pitch Rates for a Wing-Loaded Aircraft

RISK MITIGATION

Once understood, theory also provides a logical starting point for reducing the risks associated with planned out-of-control flight. At steeper attitudes (lower AOA), the airflow across the rudder has its greatest chord-wise component velocity, thereby generating greater rudder effectiveness. At the flatter attitudes expected with aft CG's and/or higher I_{yy}'s, the relative wind component is more along the vertical axis (span) of the rudder, which reduces its effectiveness. A sound build-up approach, then, would allow exploration of spin characteristics toward the edges of the intended loading envelope – without ever lacking sufficient rudder effectiveness to recover. This can be accomplished by working CG from forward to aft, and I_{yy} from minimum to maximum at each CG.

Figure 9 presents the DG-1000S CG and inertial loading envelope used for test planning. This envelope was constrained by the manufacture's certified maximum and minimum individual crew weight limits, maximum cockpit load, aircraft gross weight limit, and certified CG limits. The largest possible, or "theoretical" loading envelope could only be attained with extremely lightweight aircrew. The "test envelope" you see in Figure 9 was

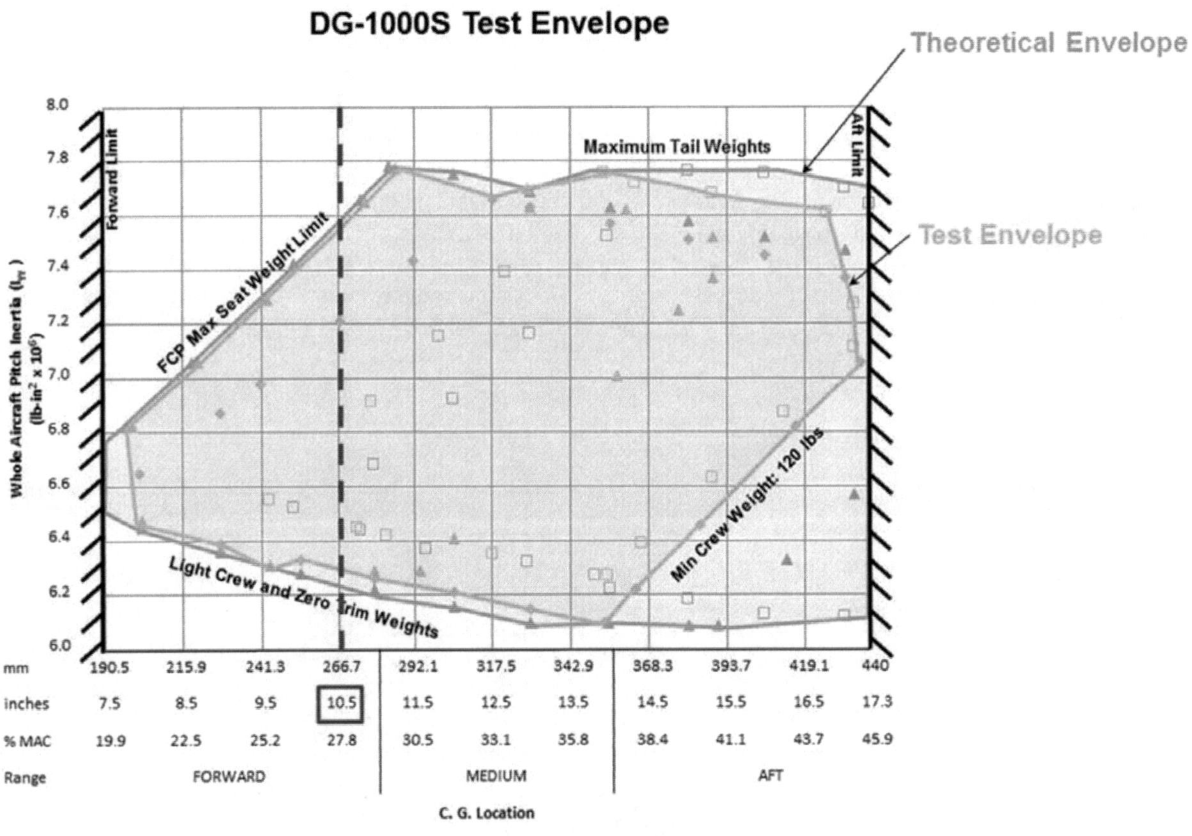

Figure 9. DG-1000S Loading Envelopes

representative of both project test pilot's (nearly identical) weight, including a parachute. With CG at/aft of approximately 13 inches, the largest and smallest possible I_{yy}'s could only be attained flying solo, with an appropriate selection of cockpit and tail ballast. At first glance, this may seem constraining on the user's requirement to conduct training with two aircrew. This was not the case, however: dual aircrew would simply operate more toward the center of the loading envelope – and often with less ballast. The test team's goal was to maximize the size of the tested loading envelope to provide the user greatest utility and safety.

A prudent starting point was to select a CG somewhat forward of where predictions suggested a stabilized spin was possible. For this project, recent spin test results on a similar aircraft – the DG-1001 – suggested starting at approximately 10 inches aft of datum, as depicted in Figure 9.[8] Next, a build-up increment of one inch was selected, working back to the certified aft limit of 17 inches. At each CG, the aircraft was first loaded to minimum-attainable I_{yy}, then maximum-attainable – essentially tracing the upper and lower edges of the test envelope seen in Figure 9. The test plan was written to allow a reduction in the size of build-up increments if early recoverability results suggested the need for a more conservative approach.

Forward CG's carry an additional type of risk for spin testing: structural failure due to over-speed/overload. After spin entry, if the aircraft's longitudinal stability is sufficient to push the nose down enough to un-stall the wing, airspeed and load factor will begin to build even if pro-spin control inputs are held. Commonly referred to as a "spiral dive," it must be recognized quickly in an extremely low-drag aircraft (like a sailplane) to avoid exceeding airspeed or load factor limits. A relatively recent SETP paper[9] presented by a sailplane test pilot – describing how he survived an in-flight break-up of his aircraft during a seemingly benign test point – underscored how easily limits could be exceeded should a maneuver not progress as planned.

Vetted test piloting skills, relevant experience and flying currency were obviously central to mitigating risk. While each project test pilot had substantial exposure to high-AoA test techniques, and decades of sailplane experience, neither had performed extensive (or recent) spin testing. This shortfall was addressed with pre-test flight training, first in an Extra-300. Training in the Extra focused on inverted spins: project pilots sought to learn how to combat the disorienting nature of such a maneuver over a prolonged period, as well as become familiar with respective susceptibility to the physiological "push-pull effect"[10] during dive pull-out – critical to preventing g-induced loss of consciousness. This was followed by acrobatics and spin training in the DG itself with an accomplished sailplane acrobatics flight instructor. Training here focused on controlling airspeed at extremely nose-low attitudes, and becoming comfortable with inverted handling qualities – essential to repeatable inverted spin entries.

Initially, calculating minimum safety altitudes for planned test maneuvers seemed straightforward. See Figure 10. Relying on previous testing of the DG-1001, the test team had a credible estimate of altitude lost during spin entry, during each turn of the spin and during any spin recovery procedure it happened to be investigating. Making the assumption that an alternate recovery procedure would fail to recover the aircraft, the pilot was given one second (or approximately 200 feet) to recognize this and transition to the flight manual recovery – which had a reported 100% success rate. Failing this, the parachute manufacturer recommended that aircrew be out of the aircraft, pulling the rip cord no lower than 1000 feet AGL to ensure full deployment before hitting the ground.

This left one question: from the time the decision to abandon the aircraft was made, to the time bailout actually occurred – how long would that take? A ground-based emergency egress test was devised: mean "time-to-bailout" for four test subjects was nearly 10 seconds – but with a large standard deviation, initially. As one would expect, repetition – developing "muscle memory" for this task – proved essential to reducing bailout time and variability. Test aircrew were required to refresh this muscle memory immediately before the first flight on every day of testing. While project pilots were eventually able to train to a bailout standard of 6 seconds or less, 10 seconds was ultimately used to leave margin for the unpredictable impact of wind blast, aircraft rotation – and adrenalin. This time period was divided by the known number of seconds per spin rotation, then multiplied by the known altitude loss per turn to arrive at a nominal altitude-lost-during-bailout figure. Terrain elevation was added to this (and the previously mentioned figures) to yield minimum test point entry altitude, minimum flight manual recovery initiation altitude, and minimum bailout altitude.

PRODUCING REPEATABLE DATA

Flight testing was divided into four phases. Phase I explored departure susceptibility using Phase A-D stalls, and the efficacy of the flight manual recovery following deliberate 2-turn spins. It also explored the widest range of CG/I_{yy} loading conditions: a "down-select" of loadings was made upon completion of this phase.

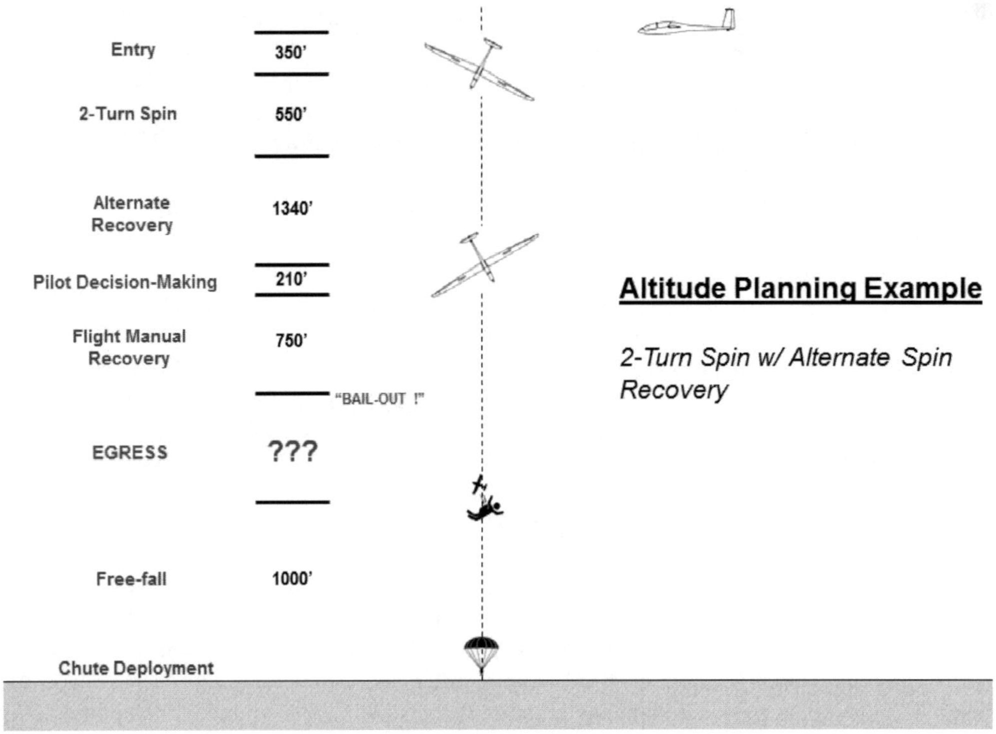

Figure 10. Planning Minimum Safety Altitudes

Only those loadings which reliably spun for two full turns (i.e. did not transition to a spiral dive) were down-selected for further examination in the remaining phases of test. Phase II was designed to explore the fully-developed spin phase during 5-turn spins. Phase III investigated aileron effects in the spin, four alternate spin recovery types, and two types of misapplied recovery controls. Phase IV investigated inverted spin characteristics during two-turn and five-turn spins. Phases I – III were also initially planned to investigate the two previously-described airspeed bleed rates for stall/spin entries.

All testing was performed in the 18m wing configuration, with landing gear and spoilers retracted. Each test point was performed three times to assess repeatability of results. Rudder inputs at the stall were attempted in both directions to check for left/right asymmetries in stall/spin characteristics.

1 kt/sec bleed-rate stall entries required a smooth, continuous increase in deck angle to the stall indication. 5 kt/sec entries could be consistently achieved with a more abrupt application of aft stick to a nearly fixed deck angle: this fixed attitude necessarily increased at aft CG's. Since lateral handling qualities tended to become "sloppy" in the DG-1000S near stall speed, careful attention to the yaw string was required for coordinated flight: anything less tended to cause the wing opposite the sideslip to drop first. Departure susceptibility data was obviously most repeatable in smooth air. Once the stall was recognized or rudder aggravation completed, all controls were abruptly neutralized; stick-fixed response was then evaluated against the pre-defined departure criteria.

Deliberate spin attempts were initiated using one of the stall entries described above. At the stall indication, ailerons were confirmed neutral, full aft stick was applied/held and full rudder abruptly applied in the desired direction of autorotation. These inputs were held for the pre-planned number of turns – or until a spiral dive was recognized. Should the latter occur, stick and rudder were smoothly neutralized while rolling to wings level, followed by a dive recovery. For spin recovery procedures, see Table 4. All began by confirming full pro-spin

Flight Manual	Misapplied Recovery Controls:	
	Incorrect Order:	**Elevator Only:**
• Full Pro-spin Rudder • Ailerons Neutral	• Full Pro-spin Rudder • Ailerons Neutral	• Full Pro-spin Rudder • Ailerons Neutral
• Rudder – Full opposite direction of spin	• Stick Steadily Forward	• Stick Steadily Forward
Pause	** Pause for 1 Second **	
• Stick – Ease forward	• Full Opposite Rudder	

Table 4. Published Spin Recovery Procedure Compared Against Misapplied Controls Investigations

rudder with neutral ailerons. The slight "pause" after applying full opposite rudder was designed to minimize yaw rate – and therefore inertial coupling – before attempting to break the stall with forward stick. (Recall Figure. 6)

While it was not practical to investigate every permutation of pilot error when attempting to recover, two types were examined. The first ("Incorrect Order"), reversed the order of the elevator and rudder input – and invited increased inertial coupling for one second. As a build-up, this procedure was always attempted first. The second misapplied controls procedure ("Elevator Only") was expected to provoke near-worst-case inertial coupling, in that pro-spin rudder was held as elevator simultaneously pushed the nose down.

As their names imply, three of the four planned "alternate" spin recovery procedures were straightforward. All controls were neutralized (and held) during the "Controls Neutral" recovery, while all controls were released and allowed to "float" for the "Hands Off" recovery. The "Rudder Only" procedure called for full-aft stick, ailerons neutral – while applying full anti-spin rudder. The recently re-named "NACA-Modified"[11] procedure required a simultaneous neutralization of elevator with full anti-spin rudder. Again, to conduct a meaningful comparison between each recovery type, all recovery inputs needed to be applied consistently: this was accomplished with abrupt, step-like inputs.

Only the Flight Manual recovery was used during recovery from inverted spins (with elevator inputs reversed, obviously).

All spin recovery attempts were considered successful and complete when rotation ceased and airspeed began to increase; a dive recovery followed.

TEST RESULTS OF INTEREST

Understanding the Data

Figure 11 provides an example time-history plot of maneuver data. If viewing in black-and-white, the slightly darker time-history traces correspond to the left-hand y-axis parameters. (In color, these appear blue) Lighter traces (or green) correspond to right-hand y-axis parameters. Pitch and roll angles/rates are body-axis values. While instrumentation was capable of measuring body-axis yaw rate, time histories present what the student would perceive during spin test training: heading and heading rate. Velocity and load factor time history plots appear in Figures 12 -15: it's important to remember that the velocity trace presents MEMs-derived inertial data: it does not depict Pitot-static airspeed. Therefore, in the following discussions, references to a fluctuation in airspeed may not be evident in the corresponding time history plot of inertial velocity. Airspeed fluctuations were reported based on post-flight review of cockpit video. Note that AHRS heading angle is displayed between $\pm$180 degrees: the transition between + 180 degrees and – 180 degrees as the aircraft rotates gives this trace a "sawtooth" appearance. Data was sampled at a rate of 10 Hz. "Entry 1" refers to a 1 kt/sec bleed-rate stall/spin entry, where "Entry 2" refers to a 5 kt/sec entry.

Phase I

The DG-1000S stall indication was consistently marked by reaching full aft stick. During the progression of Phase A-D stalls using Entry 1 – following the described CG/I_{yy} buildup – departures were first observed at 13 inches aft-of-datum during Phase C stalls. I_{yy} did not affect departure susceptibility as much as first anticipated. With Figure 11 as an example, heading rates remained steady after neutralizing controls long enough for departure criteria to be exceeded (in this case, heading change > 90 degrees) with CG's between 13 – 15 inches. This rendered the aircraft "resistant" to departure in this CG range, and "extremely resistant" at all other tested loadings. Entry 2, however, did not provoke a departure until deliberate spins were attempted; in other words, the aircraft was rated

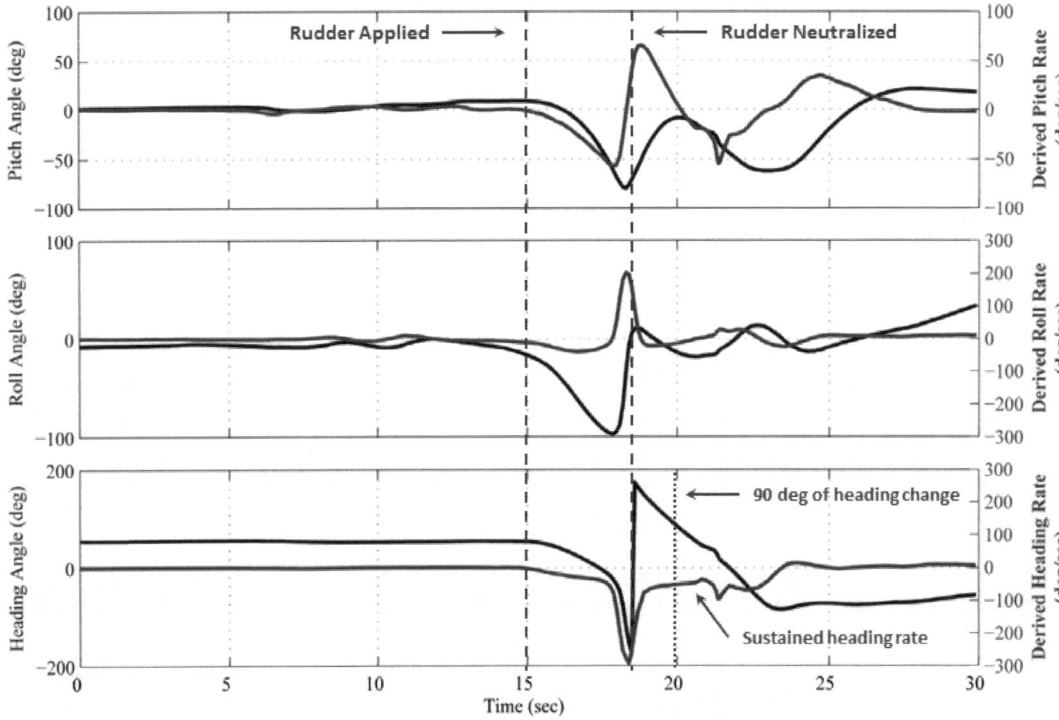

Figure 11. Departure Following a Phase C Stall

"extremely resistant" to departures at all tested loading conditions using this entry. The entry's higher pitch attitude led to substantially reduced airspeed at the stall indication – leading to reduced rudder effectiveness at the stall. Stated another way, the Phase C stall's three seconds of rudder aggravation wasn't as "aggravating" using the 5 kt/sec entry.

The impact of stall entry type can be seen more clearly in Figure 12. It provides a direct comparison of aircraft response during two-turn spins – at the same loading conditions – for each entry type. While the two spins depicted occur in opposite directions, this is inconsequential: upright left-hand spin characteristics were ultimately found to be identical to right-hand upright spins. The rudder's reduced effectiveness at the stall is clearly seen for Entry 2, in that heading rate development occurs nearly 3 seconds later than that seen during Entry 1. Another key difference is seen when comparing each spin after 180 degrees of rotation: as the aircraft pitches down to nearly vertical, the 1 kt/sec entry leads to a momentary "spike" in load factor (~2 g's), where the 5 kt/sec entry oscillates at an accelerated condition for nearly 3 seconds. While the nose pitches back up at this point for both entry types,

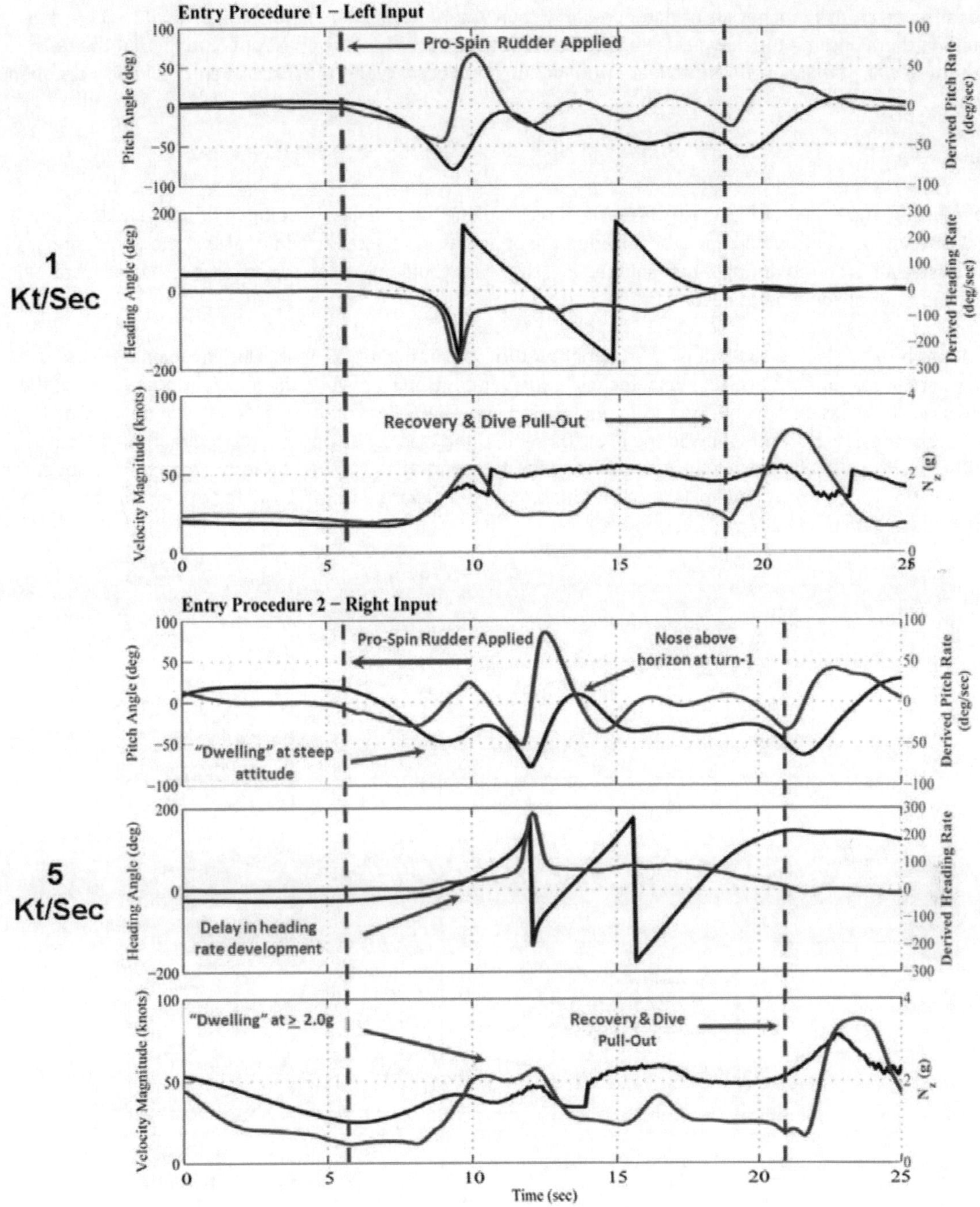

Figure 12. The Impact of Stall Entry on 2-Turn Spin Response, Identical Loading Conditions

the 5 kt/sec entry often led to much more dynamic pitch-ups – reaching attitudes as high as 20 degrees above the horizon at completion of turn-1.

As one might expect, then, more dynamic stall/spin entries led to more dynamic incipient spin phases. What was unexpected, however, was the response seen during turn-2: both entries show essentially the same oscillatory behavior, where the 1 kt/sec entry was expected to settle more quickly to a constant pitch attitude and heading rate – i.e. settle more quickly toward a fully-developed spin phase.

Only loadings at/aft of 12 inches aft-of-datum reliably spun without transitioning to spiral dives: 12 – 17 inches aft-of-datum, then, became the down-selected range of CG's for the remaining phases of test. In that the test team felt there was nothing useful left to learn from various stall entry types, the 5 kt/sec entry was also dropped after Phase I.

Phase II

As the test team moved in to this phase of testing, it was looking for at least a developed (if not fully-developed) spin: without this, students would never be afforded the opportunity to assess traditional spin mode modifiers. Because this never occurred during 2-turn spin testing, the team could only hope to converge to near-constant attitudes and rates at some point beyond the second turn.

It was not to be – see Figure 13. It depicts an upright 5-turn spin at the aft CG limit. Upright spins were assessed as "highly oscillatory" by both project pilots, technically never developing beyond the incipient phase. While the period of oscillation was 8-9 seconds, the rotational period was only 5-7 seconds: this meant that on a strict "turn for turn" basis, the aircraft rarely showed the pilot the same thing twice. The opposing aerodynamic and inertial forces (discussed previously) never achieved balance at tested conditions. Across the tested loading envelope, the attitude modifier varied from "flat" to "extremely steep," while rate varied from "slow" to "extremely rapid:" no single attitude or rate modifier could ever be assigned.

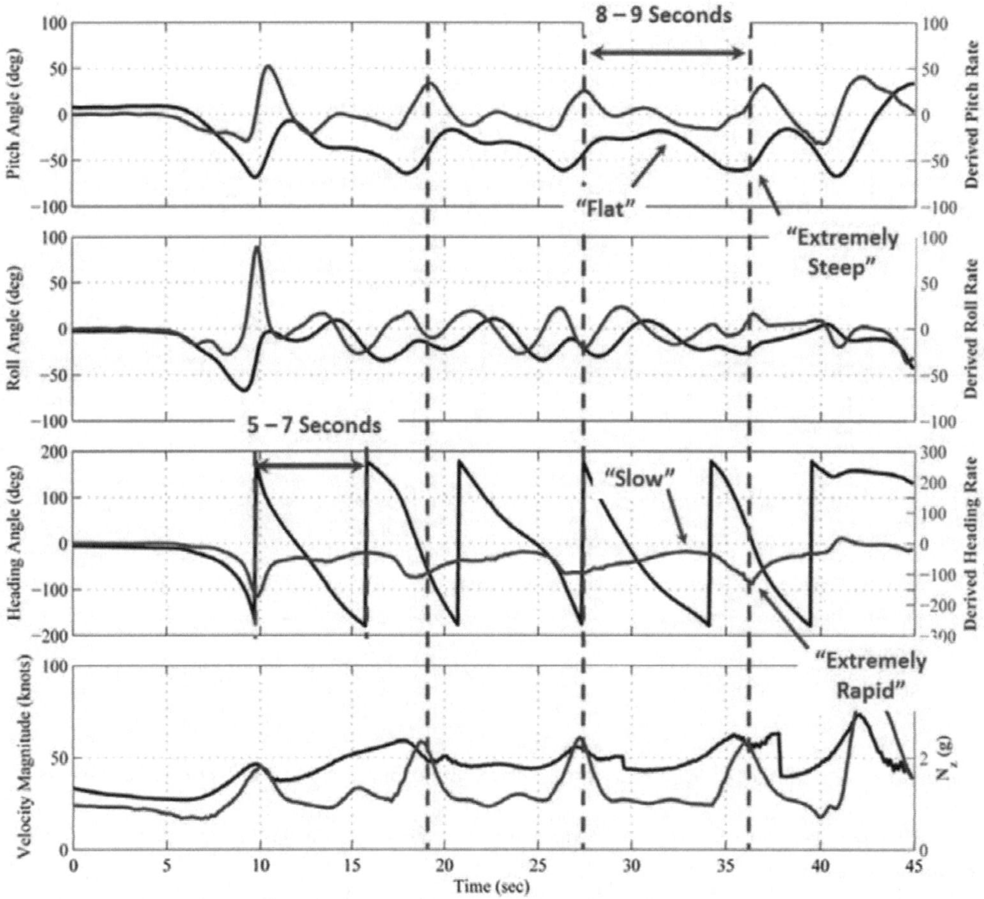

Figure 13. 5-Turn Upright Spin: Oscillation vs. Rotation Period

A closer look at Figure 13 shows that load factor spikes occurred when pitch attitude was at its steepest: combined with the increasing airspeed that was observed at that moment, the indications were more consistent with a spiral dive – not a spin. If pro-spin inputs were maintained, what followed was described by project pilots as a "snap roll on a vertical down line:" roll/yaw rates sharply increased at peak load factor, "throwing" the nose back up toward the horizon. Load factor then dropped back to 1.0, and airspeed diminished: indications consistent with a spin, not a spiral. As Figure 13 illustrates, this cycle of "spiral-like" and "spin-like" behavior continued for as long pro-spin inputs were held.

If a spin recovery was attempted during the spiral-like portion of an upright spin, when heading rate was at its highest, another unexpected dynamic emerged: rudder-lock. It announced itself with noticeably increased pedal forces while attempting to apply opposite rudder for spin recovery. Recall that rudder lock occurs following vertical fin stall: the local relative wind exceeds the vertical fin's critical angle of attack. This leads to flow separation, and can create a pressure gradient sufficient to hold the rudder against the stop – in this case, in the pro-spin direction.

This was cause for some concern. Conducting a valid comparison between various spin recovery procedures called for abrupt application of opposite rudder. Given the heavy pedal forces, the obvious concern was for the dynamic load on the aft fuselage, vertical tail and/or rudder hinges. With the center of pressure of the vertical tail a few feet above the fuselage centerline – where the fuselage was only several inches in diameter where it joined to the tail – there was concern over exceeding the ultimate torsional load limit at that location and (quite literally) twisting the tail off of the aircraft.

The risk was not merely hypothetical: abrupt rudder inputs immediately following transition from a spin to a spiral dive caused this exact structural failure mode during a 2003 sailplane spin certification test flight. In the wake of this mishap, the Federal German Accident Investigation Bureau made a formal recommendation to the European Aviation Safety Agency (EASA) to "incorporate maximum possible aerodynamic loads resulting from a combination of rudder deflection and yawing condition into the certification specifications for designs of vertical fins for sailplanes. . ."[12] The problem: this recommendation was published in 2009. The DG-1000 was certified for spins in 2001.

While all test points were performed within the certified limits of the aircraft, prudence dictated a modified approach to spin recovery in the presence of rudder lock. Either: 1) opposite rudder was applied smoothly and slowly – to minimize dynamic loading – or, 2) abrupt application of opposite rudder was delayed approximately ½-turn to allow the nose to pitch back up into the "spin-like" portion of the oscillation cycle – where no rudder lock was observed.

Phase III

All evaluated spin recovery procedures – even those involving misapplied controls – ultimately proved highly effective. As expected, the "Elevator Only" misapplied controls procedure provoked worst-case recovery characteristics, due to inertial coupling. Yet even here, autorotation only continued for one full turn after forward elevator was applied.

The magnitude of the inertial response to aileron inputs during a spin was notably affected by the timing of the input: heading rate was most visibly increased with pro-spin aileron while the aircraft was in or near the "spiral-like" portion of the oscillation cycle.

Interestingly, both the aileron effects demonstration and hands-off recovery usually left the aircraft in a high-sideslip condition at cessation of autorotation – with rudder lock. Again, to minimize tail loads, rudder was smoothly neutralized at this point to eliminate sideslip prior to dive pull-out.

Phase IV

Inverted spins were initiated from stable, coordinated, un-accelerated (-1g) inverted flight. Only 1 kt/sec entries were used. Due to wing camber, inverted spin entries occurred 20 – 25 KIAS faster than upright entries. Both 2-turn and 5-turn spins were performed at down-selected loading conditions: this meant that inverted spin testing was commenced at a CG which was 3 inches further aft than where upright testing began. This was per the planned test build-up, as Phase I testing suggested that an inverted spiral dive – not recoverability – was the greater probable threat to safety at CG's forward of 13 inches.

Figure 14 depicts the primary inverted spin mode. It too was rated "highly oscillatory," with load factor spikes reaching nearly -3.0g at the height of the "spiral-like" portion of the oscillation cycle. Unlike the upright mode, however, the oscillation period was less than the rotational period – approximately two cycles per rotation. (See the derived pitch rate trace in Figure 14) The attitude modifier varied from "Steep" to "Extremely Steep."

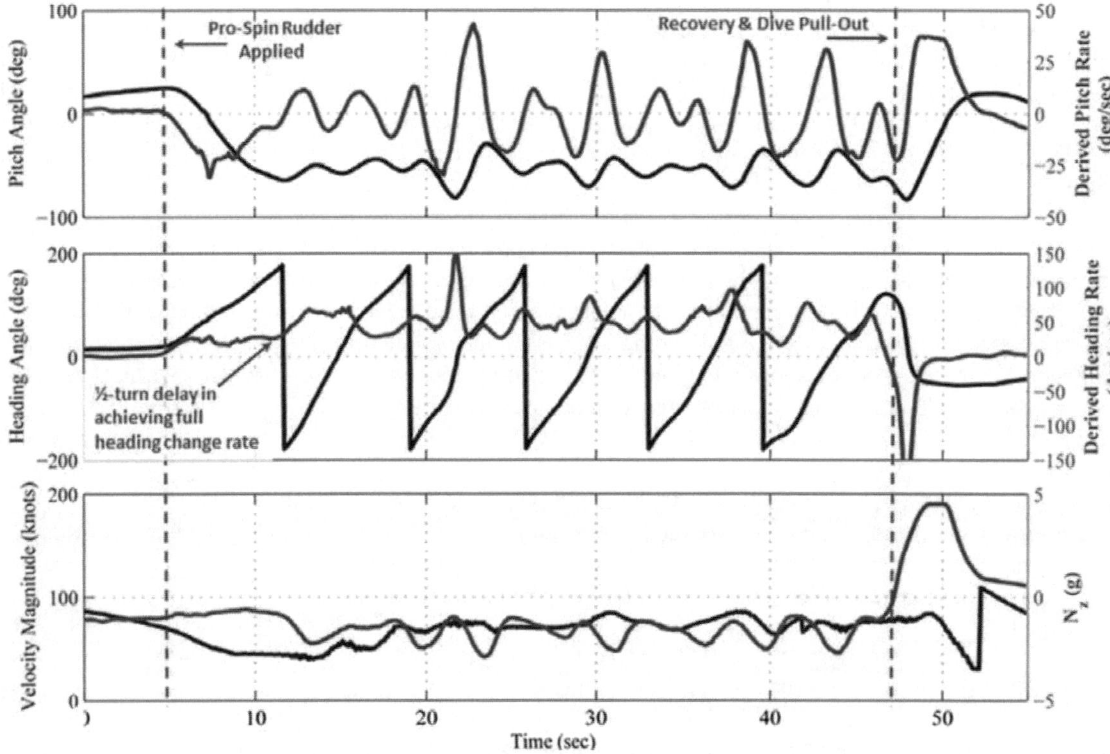

Figure 14. 5-Turn Oscillatory Inverted Spin Mode – to the Left

As one might expect, the mode was both disorienting and uncomfortable. The magnitude of the negative-g "spikes" were largest at forward-most CG's: increased longitudinal stability put the nose down slightly lower and longer, resulting in increased airspeed just before the recurring accelerated stall. While the aircraft was certified to -5.0g, negative load factors in excess of -3.0g were considered unsuitable for training – and therefore not worth the risks inherent to testing them. G-spikes at forward CG's posed a problem for the planned test build-up approach – which will be further addressed shortly.

A large majority of the inverted spins flown behaved as depicted in Figure 14. But one day, without warning – and at precisely the same loading conditions – an entirely different spin mode surfaced. See Figure 15.

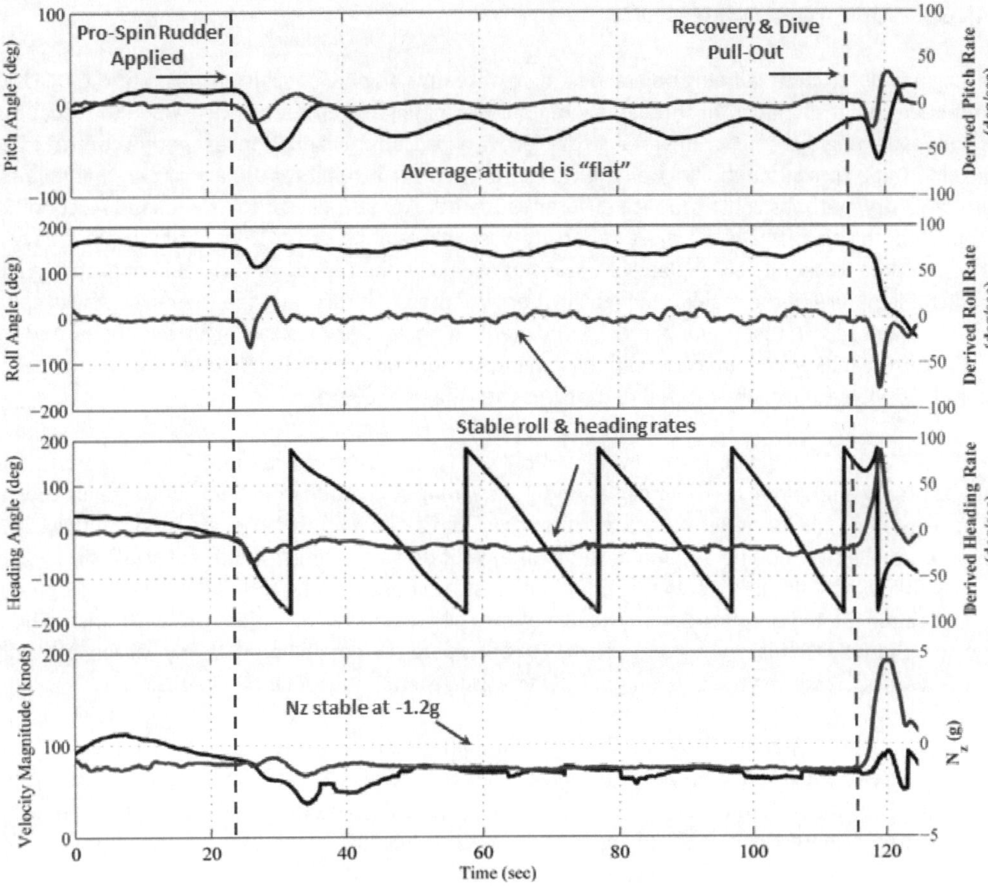

Figure 15. 5-Turn Smooth Inverted Spin Mode – to the Right (note loading is identical to Fig. 14)

This was – by far – the smoothest spin mode encountered during the program: note that pitch/roll/heading rates, as well as the load factor trace are nearly straight lines. Rate was extraordinarily slow: one rotation required 18 seconds to complete, on average. Inverted 5-turn spins lasted for over 1 ½ minutes. This was also the flattest spin observed during the program. Even stranger, it was later discovered that the oscillatory inverted mode could transition to this smooth mode – but never the reverse.

Unexpected inverted behaviors did not end there: two additional smooth modes emerged. Unlike the mode depicted in Figure 15, however, these were "Steep," with a rotational period of seven or eleven seconds, respectively. All three smooth modes had one thing in common: they only occurred when spinning to the right. After many attempts, project pilots were unable to provoke a smooth mode with left rudder applied at the stall. Smooth modes also only occurred near the aft CG limit (16-17 inches). In other words, while attempting right-hand inverted spins in this CG range, the test team was unable to reliably reproduce any one of four possible modes at will.

These unexpected spin modes are not likely attributable to an asymmetry in mass distribution: such an inertial asymmetry would likely surface in upright spin test results as well. As previously mentioned, it was not viable for the test team to modify the aircraft to gather the external air data which would likely explain these results – nor was it considered in-scope for the project.

RECONCILING RESULTS TO REQUIREMENTS

Of the five original TPS student learning objectives, the test team's original conclusion was that the DG-1000S proved an effective spin test trainer for three of them: 1) Determining the aircraft's MIL-F-83691B susceptibility to departure, 2) Investigating the efficacy of various spin recovery procedures and 3) Investigating control effects while in the spin. Departure susceptibility training was a potential strong-suit for the aircraft, in that it illustrated how various stall entry methods could produce different susceptibility ratings – at the same loading conditions. As previously reported, the aircraft was not most departure-susceptible at aft-most CG's, as classroom theory might suggest: while the data to definitively explain why was not available, the test team saw value in exposing students to the unpredictable nature of the real-world flight test environment. In this vein, the team also saw excellent learning value in inverted spin test training: the unpredictable emergence of various inverted modes, and the opportunity to learn to make accurate observations in a much more disorientating flight environment could better-prepare future test pilots for the unexpected departures they might encounter.

Per the strict MIL-F-83691B definitions, achieving two of the TPS desired learning objectives appeared problematic: 1) Identifying spin phase boundaries and 2) Assigning spin mode modifiers. Because attitude and rate modifiers were never the same on a strict "turn for turn" basis, the aircraft never attained the fully-developed phase. Moreover, these two modifiers spanned the full range of defined ratings. A student could simply guess at a particular rating following a simulated spin test point, and – at least once during the maneuver – their guess would be correct. This called the team's attention to the last of the four spin mode modifiers: "Oscillations." The "Highly Oscillatory" and "Violently Oscillatory" ratings seemed to anticipate the possibility of widely varying attitude and rate modifiers – without resolving the dilemma of how to assign them under such conditions.

All of this led the team to begin to question whether the aircraft was deficient as a spin test trainer, or if the traditional means of classifying spin behavior was, in fact, deficient. The rationale was simple: a robust scheme of classification should be able to capture the full range of observed behaviors. With the DG-1000S seeming to defy classification, a broader, more inclusive set of conventions was proposed:

First, if a range of observed attitudes and rates are bounded and unchanging from one recurring oscillation cycle to another (as Figure 13 clearly shows) -- i.e. "cycle to cycle," not necessarily "turn to turn," as the MILSTD specifies -- might it be reasonable to also call that a "fully developed" spin? If it were, the team proposed that the "developed" phase could simply be marked by the moment the axis of rotation became perpendicular to the earth, and the fully-developed phase boundary could be marked at the moment the period of the recurring oscillation cycle could be identified. Then, if the oscillations assessment is "highly" or "violently" oscillatory, attitude and rate modifiers would not be assigned: instead, the observed ranges of attitudes and rates would be reported. Such a classification scheme would allow a full description of observed DG-1000S spin mode characteristics without significantly changing how TPS challenges or develops a new test pilot's observational skills.

Within this construct, the DG-1000S begins to look more viable as a spin test training aircraft – but with two qualifiers. First, dynamic ranges of attitudes and rates are initially more difficult to discern than fixed values. Second, real-time recognition of the need to modify spin recovery technique to minimize tail loads consumes additional cognitive "bandwidth" – in a flight environment which is notorious for producing sensory overload. For these reasons, the DG-1000S may not be an optimal aircraft in which to begin spin test training: learning may be better served by exposing students to less dynamic spin modes first. At this writing, TPS is considering options to achieve these ends.

LESSONS LEARNED

In addition to implicit lessons-learned presented thus far, we feel the following learning points deserve emphasis. While a few of them are admittedly philosophical in nature, <u>all</u> are derived from *specific* events that occurred before, during or well-after testing was complete.

1. **Emergency Bailout Considerations.** We were somewhat surprised by how long it took to get out of the aircraft during our ground bailout test: our initial guess for bailout time from the DG-1000S was 5 seconds. Ground test results led us to double that figure. Don't *think* you know how long it takes to get out of your aircraft: know – as best as you practically can. We eventually reduced average bailout time by approximately 50% – through a considerable amount of practice. Published SETP papers underscore the importance of this kind of hands-on rehearsal before exposing yourself to a potential real-world bailout situation.[13]

2. **Pre-Test Training.** The primary project pilot had 20 years of test experience and nearly 35 years of experience as a licensed glider pilot: it would have been very easy to dismiss the need for pre-test glider acrobatics and spin training before commencing test. Yet the training we accomplished was the single most important step we took to mitigate safety risk: it enabled us to remain oriented and aware of what the aircraft was doing under significantly disorienting conditions, leaving us spare cognitive "bandwidth" to recognize subtleties – like the differences between "spin-like" and "spiral-like" behaviors, or the need to adjust the phasing of rudder inputs to minimize tail loads.

3. **Test Configuration Control.** As mentioned in the introduction, past SETP papers have emphasized the impact seemingly small aberrations in the aircraft mold line can have on high-AoA characteristics. While isolating and controlling variables is central to any effective test, we suspected these types of aberrations could prove especially problematic while testing an aircraft as aerodynamically efficient as a high-performance sailplane. In this regard, the DG-1000 did not disappoint.

   We began with a control-rigging check. While all control surfaces were rigged within manufacturer's tolerances, variations were noted. Two post-production modifications near the nose of the aircraft – a nose skid and a canopy repair bracket – were also documented. The main landing gear doors were held closed in flight with a simple bungee system – which was not especially effective at high sideslip angles. The relative wind would snap one door open (intermittently) during spins. See Figure 18. While the doors were relatively small and near the aircraft CG, we confirmed proper installation of the bungee system nevertheless – and ultimately concluded that the unsettlingly-loud cycling of gear doors during a spin constituted production-representative behavior.

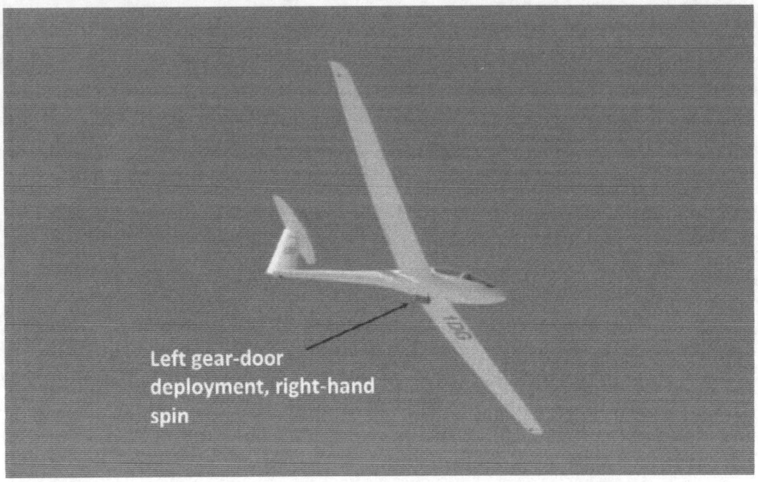

Figure 16. Intermittent Landing Gear Door Cycling

Unexplained lateral biases surfaced during early departure-susceptibility testing. One day the aircraft would tend to roll-off to the left at the stall indication; the next day to the right – at identical loading conditions. While even the lightest turbulence can cause this in a sailplane, the biases were still noted in perfectly smooth air, shortly after sunrise. Two possible causes were identified. Chord-wise gaps – several millimeters wide –

exist at each wing root, at the location where each removable wing is attached to the fuselage. Left uncovered, these gaps would allow higher-pressure air from the lower wing surface to blow the upper surface, causing substantially large flow-separation "bubbles" near the wing root. While these gaps are normally closed with tape, it becomes brittle and splits over time: when this happens asymmetrically, the wing lift distribution loses its symmetry as well. Secondly, dirt and smashed insects are also known to induce asymmetric separation bubbles on a sailplane's wings. By verifying air-tight wing-root tape and carefully cleaning the wing surfaces each morning, the previously-observed lateral biases virtually vanished.

While admittedly speculative, we believe one or more of the factors discussed in this section may have played a role in our unexpected test results. On the assumption that flow visualization tools – like tufting – do not significantly alter high-AoA aerodynamics, finding explanations for these results may be possible. Given our test training requirement, however, spending the time and money to gather such data – interesting as it no-doubt would be – was ultimately deemed unnecessary: the primary focus of the planned curriculum event was *execution* and *observation.*

Here are the lessons-learned: 1) at the very least, our successful elimination of lateral bias during departure susceptibility testing confirmed past SETP lessons-learned: minor mold line aberrations can indeed have notable impact on high AoA test results. 2) Tailoring your test to user requirements means that you must effectively identify, isolate, control and/or measure those variables that affect the achievement of test objectives – nothing more or less. While this may seem obvious, recent SETP papers[14] suggest that testers and program managers alike struggle in this area.

4. **Selecting a Build-Up Approach.** The progression of test conditions we used – from forward-to-aft CG and minimum-to-maximum I_{yy} – is a common approach to spin testing. Recoverability is the obvious driver. But the more forward the CG, the more "spiral-like" both upright and inverted spins became, where "spiral-like" behaviors led to rudder lock and structural concerns for the vertical tail. Negative load factor "spikes" during inverted spin testing were also greatest at forward-most CG's – where we had 2g's less margin available before exceeding certified limits. With a recent SETP paper illustrating how easily glider test pilots could inadvertently exceed ultimate load limits – and realizing that some structural failure modes could place the aircraft in a condition that was difficult (if not impossible) to bailout from – we stood-down from testing on multiple occasions to consider modifying our build-up. While preliminary test results strongly suggested that recoverability would not be a significant factor for this aircraft, modifying a foundational, formally-reviewed and approved safety build-up – in the middle of a test program – was no trivial matter.

   Spin mode prediction tools – assuming any are practically available to you – may not correctly identify worst-case test conditions. A recent SETP spin paper describes a test team selecting what they believed would be a benign test configuration for a start-point in their build-up – only to ultimately learn it was one of the worst-case configurations.[15] The learning points here: 1) spin recoverability at your aft-most-CG may not, in fact, represent your most serious test hazard, and 2) at the risk of sounding trite, build flexibility in to your safety buildup – commensurate with uncertainty in pre-test predictions.

5. **Testing Certified Aircraft.** As the USAF learned from the 1989 Precourt spin test, published flight manuals for certified aircraft may not tell the complete story. There was absolutely nothing in the DG-1000 flight manual to suggest we would encounter multiple inverted spin modes, for example. In that none of the inverted modes proved unrecoverable, the manufacturer may not have felt them worthy of comment, given the anticipated needs of most users – assuming the manufacturer even observed more than one inverted mode.

   It's safe to assume that JAA certification criteria did not contemplate an aircraft like the DG-1000S being used for *flight test* training. For example, the non-flight-manual spin recoveries we investigated were not evaluated during certification testing.[16] If you encounter a plan to use an aircraft in such a fashion – and *don't* hear about a commensurate plan to test – we respectfully submit it's time to *speak up*: historically, the cause of aerospace development is not well-served when placed in the hands of an un-vetted, un-trained or unwitting test pilot. If this were not so, our test pilot schools could have closed their doors long ago. As long as there are aerospace vehicles, experimental flight test will remain relevant: in the experience of this test team, ensuring others understand this may be *another matter entirely*.

6. **The Dilemma of Classification.** Before you dismiss us as apostate for daring to propose a redefinition of decades-old MILSTD spin classification conventions – so that they might accommodate our test results – it's worth remembering that nature has a decidedly inconvenient way of defying any classification scheme we may invent. An alternative approach would have been to simply conclude that the DG-1000S never really spins at all: it merely *starts to* – before coming partially un-stalled. But given that load factor and airspeed only momentarily increase during the "spiral-like" portion of the oscillation cycle, we ultimately decided that this was not an accurate characterization.

   Again, the most useful classification scheme is one which can characterize the full gamut of possible outcomes – and is standardized throughout the using community, so that we can communicate with each other unambiguously. We submit, then, that the formal cancellation of MILSTD's like MIL-F-83691B was not helpful to the cause of standardization. In their absence, professional organizations like SETP, SFTE and AIAA can help meet the pressing need to propagate updated classification conventions when the need arises, thereby restoring some degree of standardization across the industry.

7. **Kelly Johnson Was Right.** TPS leadership assembled a very small team for this effort. In 21 flying days, that team executed 147 test missions and over 800 test points – with a ground and flight test crew rarely larger than four people on any given day. This was well outside the norm for most of us, and led to some soul-searching when we first contemplated the elevated-risk portions of the program. With fewer people on our team, we knew there would be less cross-checking of each other's work within their respective specialties . . . more opportunities for errors of omission to go undetected. An unspoken, yet recurring question was: "Do we have the right people for this effort, and . . . *am I one of them*?"

   Many are familiar with the story of Kelly Johnson, and his ingenious development efforts that led to aircraft like the P-38, P-80, U-2 and SR-71. Perhaps not as well-known are his "Fourteen Rules and Practices" for such development efforts.[17] Rule #3 reads:

   "The number of people having any connection with the project must be restricted in an almost vicious manner."

   His rationale is not hard to understand: smaller teams give each member better visibility on each other's work – hence, better overall awareness of all aspects of the program. Communication is simplified and there is a greater chance for close team cohesion to develop.[18] But rule #3 does not end there – Mr. Johnson adds:

   *"Use a small number of **good** people (10% to 25% compared to the so-called normal systems)."*

   Given the inherent reduction in checks and balances of a small team, the operative word here is "good." While that word may mean different things to different people, it necessarily takes on a specific meaning within the context of elevated-risk flight testing. Independent of the other criteria we may feel beholden to, we must ask ourselves this: "Of all available choices, is this prospective team member the most *capable* – and most *willing, independent of personal self-interest* – to do what is required to help the team succeed?" Stated another way, we humbly submit that the unforgiving nature of flight test does **not** afford us the indulgence of doing anything less than putting our best-possible foot forward. While you may not always be in a position to determine the size or composition of your test team, you certainly have the ability to respectfully remind program management of proven best-practices. Our experience with this project suggests that doing less would have been dismissive of the dangers we faced. This leads to our final – and perhaps most important – lesson-learned.

8. **Defeating Dismissiveness.** For reasons we do not pretend to fully understand, say the word "glider" or "sailplane" to a flight tester . . . and in many cases, prepare for a prominent display of dismissiveness. We

were not immune: this tendency haunted us throughout the program. Yet, apart from power effects, our experience with the DG re-confirmed that a sailplane illustrates the most important aspects of spin testing – including the potential hazards. At least in part, this is why TPS has trained with them for over 20 years.

We found the dangers of dismissiveness to be insidious: generally, no *individual* who wants to be taken seriously within engineering circles overtly asserts that our attentions should vary based on whether or not the test item happens to have an engine, or by the pound of aircraft gross weight, knot of airspeed or dollar-cost. Rather, the effects of dismissiveness often seem veiled within a team's *collective* decision making.

Dismissiveness can nullify every bit of careful planning, knowledge and skill your team can muster. In spite of its insidious and potentially lethal nature, however, we submit it can be defeated. Based upon the specific forms of dismissiveness we encountered, we close with the following:

a. <u>YOU are a "Keeper of the Flame."</u> As new technologies emerge, the list of flight test topics we must understand necessarily grows – each discipline competing for our attention. As the field becomes increasingly crowded, some seem tempted to dismiss the lesser-seen disciplines. Yet there is a false premise at work here: many topics within the field of flight test are rarely seen . . . shall we dismiss them all? We submit that the relevance of a particular test discipline is not exclusively measured by its frequency-of-use, but also by the degree to which it measures *foundational airworthiness*. The need for spin testing shows no sign of vanishing any time soon: someone is going to have to do it. Keeping in mind the inherent risks, ask yourself: if not us, *then who?*

b. <u>Beware the seduction of novelty.</u> We are often engaged in exploring and evaluating technologies which have not been employed before: that which is "new" necessarily draws our attention. Spin testing, on the other hand, is nearly as old as human flight itself. Herein lies a paradox: with less than 2% of SETP papers addressing spin testing in the past 10 years, it's a safe bet that our program would feel like a very "new" kind of testing for the vast majority of active testers today. Based on personal experience, we can assure you: the fact that spin testing has been around for over 100 years will be of little comfort the moment you initiate your first elevated-risk spin test point. Like frequency-of-use, we submit the importance of a test discipline should not be assessed by how "new" it may be.

c. <u>Realize there are things worse than confrontation.</u> Anyone of us who have had to bury a fellow tester probably appreciates this best. Preventing disaster requires you to speak up when you encounter a hazard that has not been adequately dealt with. It's critical to have your facts correct beforehand, and worth the time to find relevant, credible third-party sources to support your position. If the engineering facts back you, persist: respectfully, tactfully – yet unapologetically – until the matter is satisfactorily resolved. No, you may not always make friends in the process . . . but understand that your silence under such circumstances <u>*absolutely counts*</u> as a 'go' vote: if it didn't, you probably wouldn't be on the test team. Are you prepared to live with the consequences?

d. <u>Publish.</u> Widely-understood test hazards are more difficult to dismiss. Spin testing is notoriously hazardous, yet perhaps not well understood, given the small number of us who have done it. This is no small irony: the dismissiveness this lack of understanding invites can stand as an obstacle to being published . . . thereby helping multiply the dismissiveness. To the credit of the SETP Publications Committee, we offer this paper into evidence: persistence *can* overcome a harmful dynamic that might otherwise prove self-sustaining. It should be clear that our test team owes a debt of gratitude to those who came before us – and who had the foresight to allow us to preemptively learn what they discovered. As you weigh the merit of following in their footsteps, also consider this: to the extent that your efforts to publish life-saving lessons-learned encourage your peers to do likewise, the life you ultimately save may be your own.

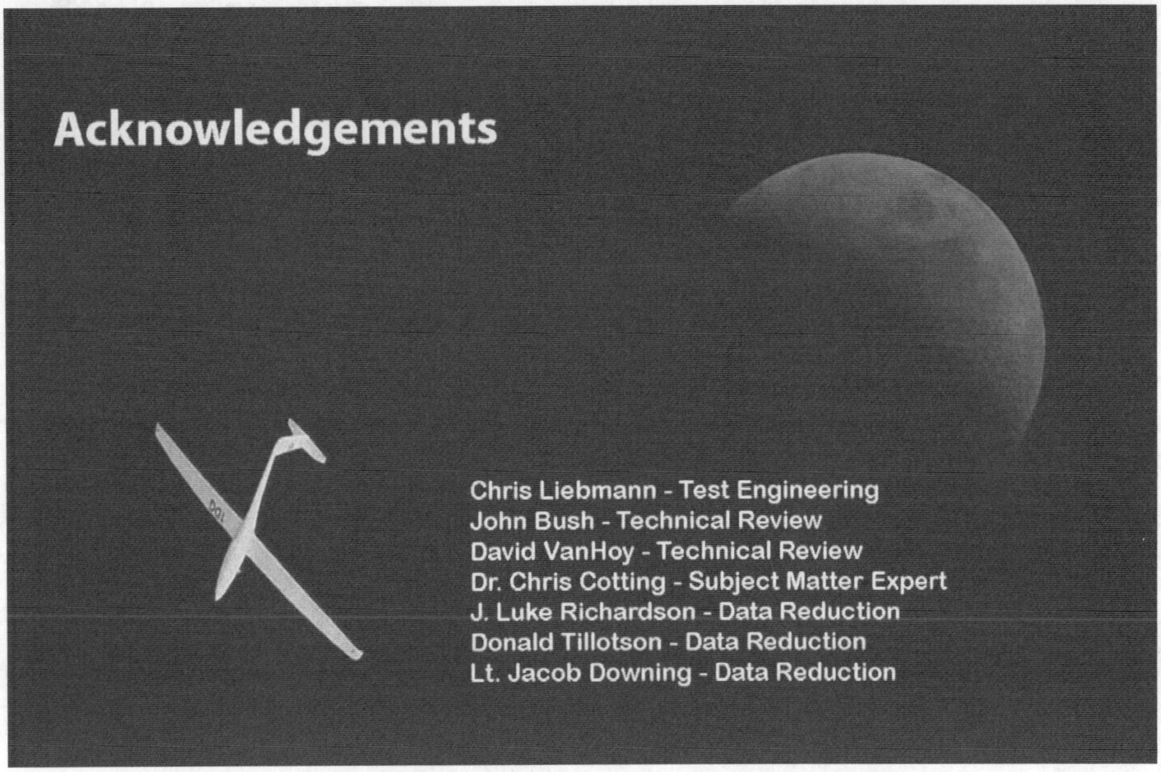

---

[1] Sonja Englert, "Lancair Columbia 400 Spin Recovery Testing," Proceedings of the 49th Annual SETP Symposium, 2005, pg.11

[2] Ibid., pg. 12

[3] LCDR Robert Bibeau, CDR Terry "Motek" Barrett, Mike Heller and Scott Hoffman, "Benign Departure: Determining Angle of Attack Limits for Flight with Lateral Weight Asymmetries," Proceedings of the 52nd Annual SETP Symposium, 2008, pg.1

[4] Mike Alsbury, Clint Nichols and Mark Stucky, "SpaceShipTwo - The Road to Rocket-Powered Flight Test," 57nd Annual SETP Symposium, 2013

[5] Doyle Janzen and Charles Precourt, AFFTC-TR-89-7, Schleicher ASK-21 Glider (TG-9) Stall and Spin Evaluation. (Air Force Flight Test Center, Edwards AFB: 1989)

[6] United States, Department of the Air Force, Military Specification Flight Test Demonstration Requirements for Departure Resistance and Post-Departure Characteristics of Piloted Airplanes – Notice of Cancellation (Edwards AFB: 1996) Pg.1

[7] MEMSIC 440 Series User Manual, Doc# 7430-0131-02 Rev. A. (Milpitas, CA: MEMSIC, Inc., 2010), pg. 2

[8] AFFTC-TR-11-66, United States Air Force Academy TG-16A Training Glider Qualification Test and Evaluation, (Air Force Test Center, Edwards AFB: 2011)

[9] David Lawrence, "Anatomy of an Accident – Flutter Induced Wing Separation in an Optionally-Piloted Ultra-Light UAV," SETP 39th San Diego Symposium, 2009

[10] NAWCADPAX-99-74-TR, Human Physiological Responses to Push-Pull Acceleration as Experienced in Helicopters, (Naval Air Warfare Center Aircraft Division, Patuxent River: 1999) Pg. 2

[11] 412TW-TP-14-18, DG-1000S STALL AND SPIN FLIGHT TEST AND EVALUATION (Air Force Test Center, Edwards AFB: 2015) Pg. D-8

[12] Federal Republic of Germany, Federal German Aircraft Accident Investigation Bureau, Investigation Report 3X221-0/03 (Braunschweig: 2009) Pg.5

[13] Tom Porteous, "Safety Considerations in Spinning Small Light Aeroplanes," Cockpit Magazine, Society of Experimental Test Pilots (Lancaster: 2006) Pgs. 7-8

[14] Lt Col Don Grove, Maj Garret Knowlan, "CV-22 Engine Inlet and Scroll Time on Wing Improvement Test," Podcast, 56th Annual SETP Symposium, 2012

[15] Elglert, Sonja, *Lancair Columbia 400 Spin Recovery Testing*, 2005, *Proceedings*, 49th SETP Symposium, pg. 12.

[16] S. Lehner, JAR 22.221 Proof of Compliance Report for DG-1000 (Bruchsal: 2001)

[17] *Kelly's 14 Rules & Practices,* http://www.lockheedmartin.com/us/aeronautics/skunkworks/14rules.html

[18] Phillips, Jeffrey, *The Smaller, the Better,* http://workingsmarter.typepad.com/my_weblog/2005/09/the_smaller_the.html

412TW-PA-16463

FOOL ME ONCE: LESSONS LEARNED FROM TESTING THE F-16 HYBRID FLIGHT CONTROL SYSTEM

Major Jameel Janjua, RCAF, Project Pilot, 416th Flight Test Squadron
Major Joseph Schenkel, USAF, Project Pilot, 40th Flight Test Squadron
Jessica Peterson, Test Director, 416th Flight Test Squadron
Jay Welch, Flight Test Engineer, Lockheed Martin

4
1
2
T
W

SEPTEMBER 2016

412TH TEST WING
EDWARDS AIR FORCE BASE, CALIFORNIA
AIR FORCE MATERIEL COMMAND
UNITED STATES AIR FORCE

ABSTRACT

In 2014, the Automatic Ground Collision Avoidance System (Auto GCAS), developed to reduce controlled-flight-into-terrain and spatial-disorientation mishaps, was implemented on 600+ USAF F-16 Block 40/50 aircraft. Since that time, four aircraft and pilots have been saved by the operational system. However, Auto GCAS is still not fielded on the 1,400+ USAF and international partner pre-Block 40 analog flight control computer F-16s. In March 2016, the 416th Flight Test Squadron at Edwards Air Force Base, California, successfully completed flight test on a modified F-16 analog flight control computer, which provides Auto GCAS capability on pre-Block 40 aircraft and incorporates improvements developed from the fielded Auto GCAS. Previously it was discovered that during high-g split-S maneuvers the fielded Auto GCAS would cause nuisance activations. Follow-on test pilot/government/contractor collaboration and simulation development resulted in an improvement that retained mishap protection while allowing normal split-S maneuvers. Although the overall system was considered mature and performed as expected during the majority of flight test, safety planning was implemented to prevent reliance on the system under test. Two instances of correctly exercised safety mitigations are discussed in this paper. During the first attempt of the fastest, steepest diving test point, pre-planned safety procedures proved critical in protecting the pilot when an unanticipated system failure occurred. On a subsequent test flight, the primary method used by the pilot to override the automatic system malfunctioned and required safety mitigations to be exercised. This paper addresses the flight test of Auto GCAS on pre-Block 40 analog F-16s, development and improvements made to allow high-g split-S maneuvers, and the successful application of safety practices after system failures.

F-16 AUTOMATIC GROUND COLLISION AVOIDANCE SYSTEM

INTRODUCTION

The Automatic Ground Collision Avoidance System (Auto GCAS) and Pilot Activated Recovery System (PARS), developed to reduce controlled-flight-into-terrain (CFIT) and spatial-disorientation mishaps, was fielded in September 2014 on USAF F-16 Block 40/50 aircraft. It was the culmination of several decades of development with the original F-16 GCAS developed on advanced fighter technology integration F-16 (AFTI/F-16) starting in 1984, with further development into the 1990s. In 2006 the U.S. Secretary of Defense mandated a 75 percent reduction in DoD mishaps. The leading cause of loss of life and aircraft in fighter fleets, CFIT and spatial disorientation, required mitigation in order to achieve reduced-mishap numbers. From 1991 to 2014, within the U.S. F-16 fleet alone, there were 51 Class A CFIT and spatial-disorientation mishaps resulting in the loss of 39 lives and 49 aircraft.

The AFTI/F-16 Auto GCAS technology was first placed on a path to becoming part of the production Block 40/50 software suite under a 2010 collaboration among the Air Force Research Laboratory (AFRL), NASA Dryden (now Armstrong) Flight Research Center, the Air Force Flight Test Center, and Lockheed-Martin Aeronautics (LM Aero) in the Fighter Risk Reduction Project (2007-2010). The technology then transitioned to the F-16 Program Office. The 416th Flight Test Squadron (416 FLTS) completed the Auto GCAS Design Tryout (2011-2012) and M6.2+ (2012-2013) test efforts, resulting in fielding of the system in September 2014 on all USAF Block 40/42/50/52 F-16 aircraft, encompassing approximately 610 digital flight control computer (DFLCC)-equipped aircraft.

Since fielding, four aircraft and pilots have been saved by the system. The first save was during a high-angle strafe operational mission in November 2014 in which the pilot was using multiple strafe passes to provide close air support (CAS) to troops in contact. The second save was in January 2015 during a high-aspect basic fighter maneuver (BFM) engagement over the sea, where a steep dive angle combined with a high-airspeed maneuver required Auto GCAS activation. The third Auto GCAS save was in August 2015 during a low-level terrain-masking training mission in which the pilot was unaware of a peak ahead of the flight path during an inverted ridge crossing with excessive dive angle. The most recent save was in May 2016 after a pilot experienced g-induced loss of consciousness (G-LOC) during a high-g training maneuver.

Though the fielded Auto GCAS on USAF Block 40/50 F-16 aircraft has been extremely successful, several deficiencies were identified once the system was fielded on the USAF Block 40/50 F-16s. Causes of these deficiencies included different hardware components, hardware failures that were not appropriately mitigated within Auto GCAS, and corruption of data files, all resulting in erroneous Auto GCAS activations. Additionally, during some high-g split-S maneuvers the fielded Auto GCAS would cause nuisance early activations. This paper will detail test pilot/government/contractor collaboration during recent years that resulted in the development, testing, and implementation of fixes to these deficiencies.

Following the successful fielding of the Auto GCAS on USAF DFLCC-equipped aircraft, the AFRL program office initiated a new phase of the Auto GCAS program to identify a technical solution that would enable integration of Auto GCAS into the analog flight control computer (FLCC) found in pre-Block-40 F-16s. LM Aero developed a system in which redundant digital processor cards were added to the analog flight control computer without affecting existing analog cards, resulting in a hybrid digital/analog flight control architecture. This architecture provided a technical path forward for incorporating advanced automated capabilities onto the 1,400+ USAF and international partner pre-Block 40 analog FLCC F-16s.

Known as the Hybrid Flight Control Computer (HFLCC) test program, this technology demonstration program was conducted over two phases and 30 test flights between March 2015 and March 2016 at the 416 FLTS at Edwards Air Force Base. The program successfully demonstrated that HFLCC preserved legacy analog F-16 flying qualities and was able to host a digital application, specifically the Auto GCAS and PARS, on analog FLCC F-16 aircraft.

OVERVIEW OF AUTO GCAS

The Auto GCAS was designed to prevent CFIT mishaps by providing an automatically initiated and executed maneuver when a collision with terrain was determined to be imminent. The PARS was designed to provide a manually initiated automatic recovery that returned the aircraft to wings-level flight in the event that a pilot recognized s/he had become spatially disoriented. Critical to the design of the Auto GCAS, the system was designed to prevent collisions while never interfering with normal mission operations or harming the pilot or aircraft. Because of extensive collision avoidance technology development using the F-16 test bed aircraft over prior decades, the F-16 fleet was chosen to be the first fighter fleet to field the Auto GCAS and PARS.

Auto GCAS Design

The Auto GCAS would determine when a collision with terrain was imminent by utilizing an aircraft trajectory prediction algorithm (TPA), based on aircraft state, navigation position, and system altitude, and comparison with a terrain profile generated from onboard digital terrain elevation data (DTED). At the instant the TPA touched the Protection Altitude (including buffers), the Auto GCAS executed an automatic recovery. The Auto GCAS recovery maneuver was an aggressive roll to the nearest horizon and a nominal 5 g pull-up. If the aircraft was already at an elevated g during an activation, Auto GCAS would perform either an unload roll-and-pull maneuver or a pull-through maneuver depending on the aircraft attitude and the pitch rate. This aspect of the design was instrumental in a lesson learned after the system was fielded and will be discussed later in this paper. The automated recovery could be terminated (paddled-off) by the pilot at any time by depressing the paddle switch on the side stick controller. For flight test only, up to 9,999 feet of additional clearance could be added to the terrain profile by using a pilot-entered terrain clearance buffer (TCB). This collision estimation and automatic recovery process is illustrated in figure 1.

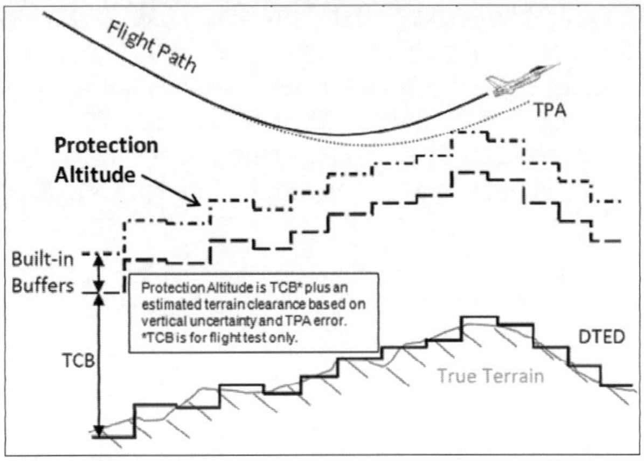

Figure 1 Auto GCAS Recovery Illustration

Since spatial disorientation had been a common cause of CFIT mishaps, the PARS was designed to provide a disoriented pilot with an automated recovery to straight and level flight from both nose-low and nose-high conditions. The pilot activated the PARS by actuating the advanced mode (ADV MODE) switch. From a nose-low attitude the PARS performed the same automated maneuver as did the Auto GCAS, returning the aircraft to level flight. From a nose-high attitude the PARS performed one of three maneuvers: 1) a roll to upright followed by a 0.5 g bunt to the horizon; 2) a roll to 90 degrees bank followed by a 0.5 g slice to the horizon and a roll back to upright; or 3) a roll to inverted followed by a 1 g inverted letdown to the horizon and a roll back to upright. The type of nose-high PARS would depend on flight path angle and calibrated airspeed at the initiation of the maneuver, as shown in figure 2.

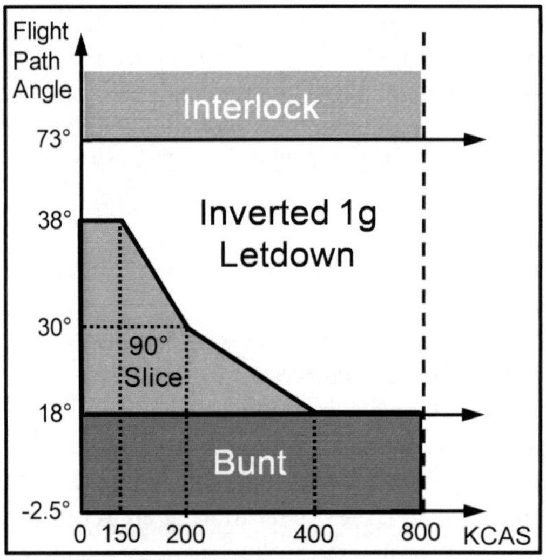

Figure 2 Nose-High PARS Recovery Modes

The Auto GCAS logic that determined the need to avoid resided in the advanced data transfer equipment (ADTE); the logic and command augmentation required to perform the automated maneuver resided in the DFLCC. No hardware modifications were required for implementation of the Auto GCAS on the Block 40/50 DFLCC-capable F-16 aircraft; however, some software updates were required. The Auto GCAS was a complex system of systems requiring numerous components to include the DFLCC, ADTE, the pilot-vehicle interface (PVI) and the navigation solution from the INS/GPS. A visual representation of hardware components required is depicted in figure 3. Implementing the Auto GCAS on pre-Block 40 aircraft with an analog FLCC would employ many of the same hardware and software components; a hardware modification, however, was required for the HFLCC, which will be discussed later in this paper.

Figure 3 Auto GCAS Aircraft Components

Operationally, the amount of margin used to determine the need to evade could be controlled by the pilot via selection among the normal (NORM), minimum (MIN), airshow (SHOW), and OFF protection modes. The NORM mode was designed to provide the largest amount of protection with nuisance-free protection for flight at and above 500 feet above ground level (AGL). The MIN mode was designed to provide nuisance-free operation with reduced collision protection for flight at and above 300 feet AGL, depending on the roughness of the terrain and the aggressiveness of pilot flown maneuvers. Additionally, two automatically selected sub-modes (STRAFE and LOWLEVEL) would tailor the amount of conservatism during strafe attacks and low-altitude flight (< 1,000 feet AGL) when in either the NORM or MIN protection mode, without requiring pilot interaction.

A scan pattern was used to determine which DTED was relevant for inclusion in the terrain profile. Figure 4 provides a generic representation of the scan pattern overlaid on a representation of a digital terrain map where each point represents a terrain elevation. The points contained within the scan pattern were used to generate the 2-D terrain profile. There were two scan pattern shapes, fan and circular, shown in figure 5. The fan shape was used for straight or turning flight, and was configured based on the Auto GCAS Mode (NORM, MIN, and SHOW), low-level mode, strafe mode, and aircraft configuration. The circular shape replaced the fan pattern when the aircraft was in a steep dive (approximately greater than 60 degrees). This scan pattern had a 3-nautical-mile radius, and the highest terrain post within the scan pattern determined the elevation of the entire terrain profile when this scan pattern was selected. Thus, higher terrain near the edge of the scan pattern could trigger a recovery.

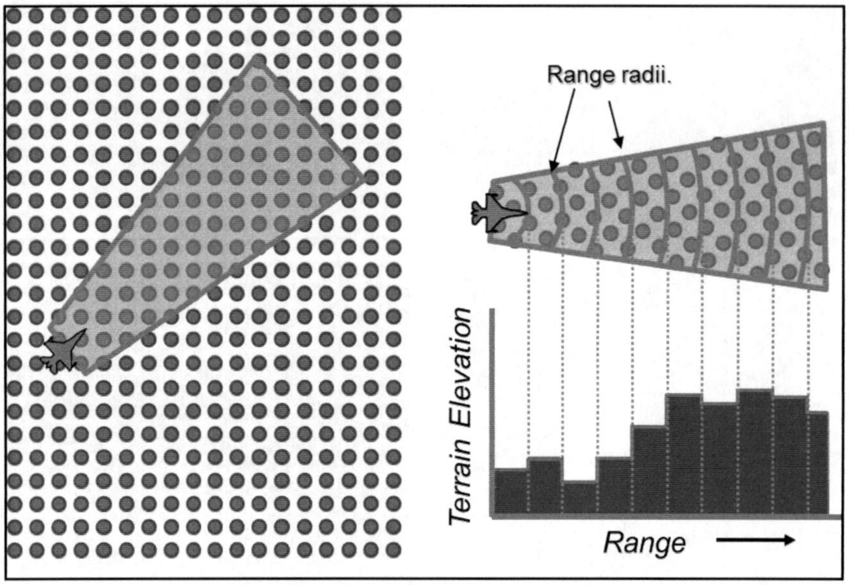

Figure 4 Auto GCAS Scan Pattern

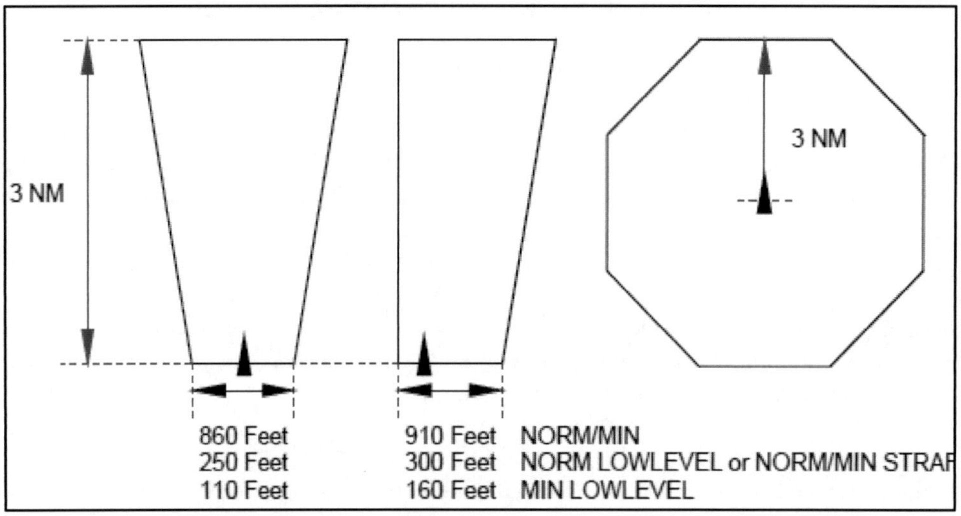

Figure 5 Auto GCAS Scan Pattern Shapes

The Auto GCAS and PARS were designed not to defeat or de-augment existing safety functionality (AOA-limiter, g-limiter, etc.). Additionally, both the Auto GCAS and PARS had software interlocks and system-wide integrity management (SWIM) that prevented system operation under conditions that could result in unsafe or unwanted system activation. The term "Terminate" referred to actions that would stop an Auto GCAS flyup or PARS recovery that was under way. "Inhibit" referred to action that prevented a flyup or recovery from occurring. Notably, introducing a failed state to Auto GCAS would not terminate a flyup in progress but would inhibit future flyups. The most glaring difference is in the paddle switch: it will terminate both an Auto GCAS flyup and PARS recovery, but holding the paddle switch only inhibit PARS recoveries. The ability of the pilot to immediately terminate Auto GCAS and take control of the aircraft was a critical design feature and related to one of the lessons learned described later in the paper.

Auto GCAS Objectives

During the multi-decade development of Auto GCAS technology, a fundamental design philosophy was adopted to guide decisions when design tradeoffs had to be made. This fundamental philosophy consisted of three parts, which in priority order were: 1) do no harm to the aircraft or pilot; 2) do not interfere with normal mission operations; and 3) prevent ground collisions. These three 'pillars of automatic collision avoidance' provided the framework around which both the system design and test approach were built. The highest priority, 'do no harm,' was focused on ensuring that automatic maneuvers did not exceed aircraft limit, cause pilot physiological issues, or damage aircraft structure. The 'do not interfere' aspect was focused on ensuring that the pilot could retain the ability to do normal F-16 missions such as low-level flight, strafe attacks, high-g split-S maneuvers, and other operations without any nuisance Auto GCAS activations. Finally, the 'prevent collisions' pillar was focused on preventing controlled-flight-into-terrain mishaps. Developmental flight test of the Auto GCAS on DFLCC Block 40/50 aircraft consisted of over 200 flights and 2,000 Auto GCAS recoveries, and 40 test pilots evaluated the performance with respect to all three design aspects.

SPLIT-S IMPROVEMENTS TO AUTO GCAS

Despite the extensive developmental testing described previously, some nuisance activations were discovered after fielding the Auto GCAS to the operational USAF F-16 aircraft. During high-g split-S maneuvers the Auto GCAS could cause nuisance activations that would interfere with BFM training missions. Nuisance Auto GCAS activations tended to occur during very high-g (>6 g) split-S type maneuvers with high power settings, typically full afterburner (A/B). Additionally the worst-case activations tended to occur when the aircraft was inverted at about 60-80 degrees nose-low. An issue with these types of nuisance Auto GCAS activations was that they would often command a roll-and-pull maneuver that consisted of reducing the g, rolling the aircraft then increasing the g back up to 5 g. During the unload maneuver, the airspeed would significantly increase and the aircraft would remain at a nose-low attitude for a longer period of time. This would result in a greater loss of altitude and larger increase in airspeed than if the aircraft had been allowed to continue a high-g pull-through maneuver. Figure 6 depicts the conditions of nuisance activations during BFM training missions, with green showing the intended pilot flown maneuver staying above the 5,000 feet AGL floor with a higher-g pull-through maneuver. The corresponding Auto GCAS nuisance roll-and-pull maneuver is shown in red with a larger altitude loss.

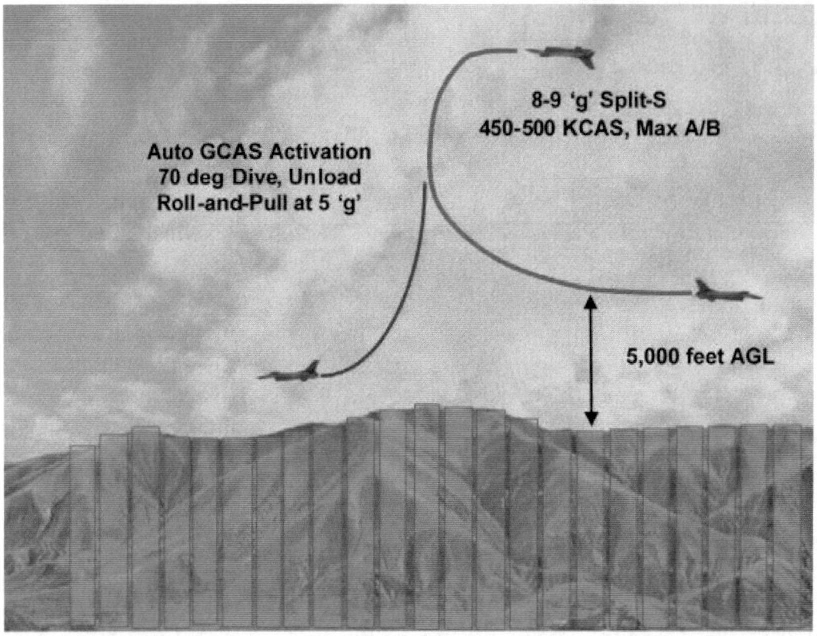

Figure 6 Auto GCAS Nuisance Activations during BFM Training Missions

Despite the BFM nuisance activations, front line F-16 squadrons choose to fly with this system active due to the benefits of Auto GCAS protection. Though there was no original scheduled opportunity to fix and test a split-S improvement to Auto GCAS, the test team developed a plan to use existing ongoing testing for the Analog Auto GCAS program on HFLCC (discussed later) to devise and test an improvement to the split-S Auto GCAS mechanization. Follow-on test pilot/government/contractor collaboration and simulation development resulted in devising an improvement that retained mishap protection while allowing normal split-S maneuvers. Two improvements were incorporated: 1) making the Auto GCAS roll-and-pull maneuver less likely, and 2) increasing the pull-through g utilized by Auto GCAS to increase protection.

The first improvement involved modifying the boundary condition, termed the decision boundary, between roll-and-pull and pull-through Auto GCAS activations. The decision boundary was governed by an equation based on aircraft attitude and pitch rate. Extensive test pilot-flown split-S maneuvers within the LM Aero simulators resulted in devising a decision boundary equation dependent on airspeed to reduce the frequency of roll-and-pull maneuvers while retaining mishap protection. Figure 7 shows a notional depiction to the change in the decision boundary, with the red dashed line showing the original boundary between roll-and-pull and pull-through maneuvers and the black dashed line showing the improved condition.

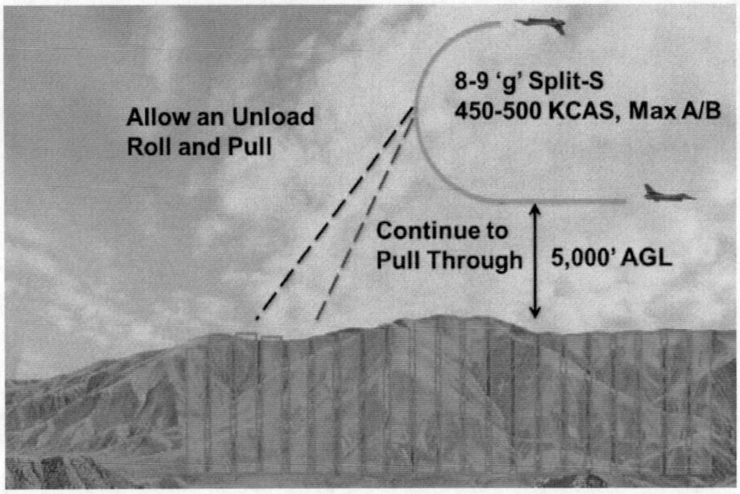

Figure 7 Improvement to Decision Boundary for Roll-and-Pull Auto GCAS Maneuvers

The second improvement involved increasing the g that the Auto GCAS algorithm would use during pull-through maneuvers. Originally, Auto GCAS was designed to pull up to 5 g's to prevent injury to the pilot and overstressing the F-16 airframe for all available loadings. This proved to be sufficient for the majority of conditions, providing protection for G-LOC scenarios while retaining nuisance-free capability for the majority of split-S maneuvers down to the 5,000 feet AGL fight floor used in training. However, some high-g (greater than 6 g) split-S maneuvers were resulting in nuisance activations. There were multiple proposals on how to improve the situation; however, for flight test the Auto GCAS was modified to continue a pull-through with a higher-g recovery. Figure 8 depicts the higher-g improvement, with the original Auto GCAS 5 g recovery shown in red and an example 6 g continued pull-through recovery shown in green.

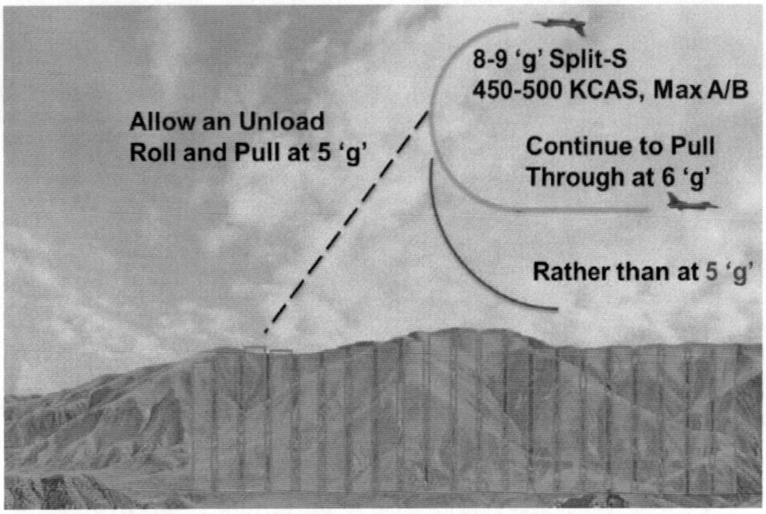

Figure 8 Increase in Auto GCAS Pull-Through G Command

Following flight test, the team returned to the LM Aero simulator facilities and further refined the boundary condition for the pull-through/roll-and-pull mechanization as well as the higher-g command pull-through mechanization. Additionally, the final solution on the automatic maneuver will need to be incorporated into the TPA prediction within the Auto GCAS algorithm. At the time this paper was written, the final mechanization has yet to be determined. Ongoing simulator investigations have proven central to the discussions of choosing the best improvements.

HYBRID FLIGHT CONTROL COMPUTER (HFLCC)

The Auto GCAS was fielded to the USAF Block 40/50 F-16s in the fall of 2014 and has saved four pilots' lives and aircraft to date. Currently, however, there remain over 1,400 pre-Block 40 F-16 aircraft with analog FLCCs that are not protected with Auto GCAS. The HFLCC provided a means of adding flight control software onto analog FLCC F-16s. The HFLCC consisted of the original analog FLCC with addition of four digital cards within existing spare slots. The new digital cards hosted software capabilities such as Auto GCAS within the outer loop of the control law, resulting in automatic capabilities without affecting legacy flying qualities.

Starting in 2015, the HFLCC flight test program demonstrated technology that incorporated a hybrid digital and analog FLCC. Between 2015 and 2016, the two-phase flight test program demonstrated that the HFLCC system preserved legacy analog FLCC F-16 flying qualities and was able to host digital applications, in particular the Auto GCAS and PARS, in analog FLCC F-16 aircraft.

HFLCC Design

The HFLCC was a modified version of the standard analog FLCC currently operated on all pre-Block 40 aircraft. Aircraft modification consisted of integrating new digital modules with existing analog modules to create a hybrid computer. The new digital modules were integrated to host advanced autopilot functions such as those required to integrate Auto GCAS and PARS. Aircraft 80-3666, figure 9, was selected as the HFLCC test bed.

Figure 9 HFLCC Test Aircraft 80-3666

Baseline Analog FLCC Design

A basic knowledge of the unmodified analog FLCC is helpful in understanding the modified HFLCC. The unmodified analog FLCC on the pre-Block 40 F-16 was a quad-redundant analog computer, shown in figure 10. The FLCC consisted of four branches, with each branch containing an independent power supply, and four analog modules, each hosting the inner-loop control laws for the aircraft. The control laws on these analog modules contained the logic that commands control surfaces on the aircraft and were responsible for the resulting flying qualities of the F-16. The FLCC also included the motherboard, which connected the analog and discrete inputs from the five wire harnesses to the analog modules in each branch. Each branch of the FLCC had one spare slot that was previously unused.

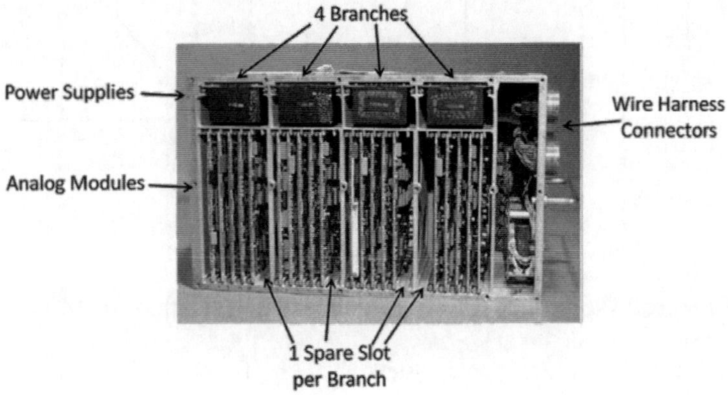

Figure 10 Unmodified Analog Flight Control Computer

HFLCC Design

The overall goals for the HFLCC design were threefold: 1) host digital applications; 2) retain legacy flying qualities of the analog FLCC; and 3) retain the safety features originally designed into the analog FLCC.

In order to host digital applications on the pre-Block 40 aircraft, the modified HFLCC integrated four new digital modules into the existing FLCC spare slots. These digital modules hosted the Auto GCAS autopilot software. Auto GCAS commands were sent to legacy analog modules by intercepting the pilot stick and rudder signals, negating the pilot commands and inserting the appropriate stick and rudder commands to perform the Auto GCAS recovery maneuver. The HFLCC was able to retain the legacy flying qualities of the analog FLCC since the modification contained no changes to the inner-loop control laws hosted on the legacy analog modules. Flight test demonstrated that the expected F-16 flying qualities were unchanged by incorporation of the digital cards.

Finally, the inherent safety features of the analog FLCC were retained within the HFLCC through a failure management system using a "fail off" philosophy on the new digital modules. To implement this philosophy, fail safe switches were added to the motherboard to prevent the digital modules from affecting the pilot signals going to the legacy analog modules in the event of a failure. Figure 11 shows the pilot signal routing through and around the digital module of a single branch of the HFLCC. The standard mode for HFLCC where pilot commands were routed through the digital cards was termed "PASSTHROUGH" mode. In the event that a failure was identified on a particular branch, that branch's digital module was set to a "BYPASS" mode. In this state pilot signals bypassed the digital modules and went straight to the legacy analog module, providing a safe mode of operation. During flight test, a safety build-up was implemented by testing and evaluating all flight test conditions in the BYPASS mode prior to PASSTHROUGH mode.

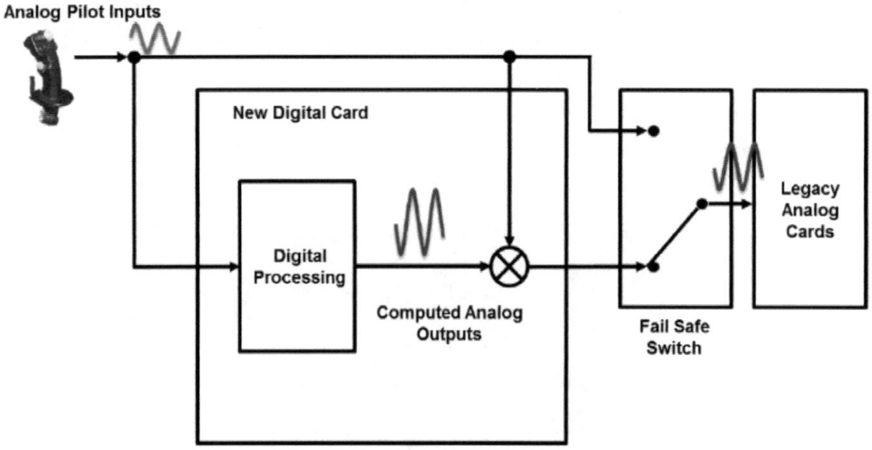

Figure 11 Pilot Signal Routing on a Single Branch of HFLCC

In addition to the HFLCC failure management system, flight test-specific design features were incorporated. Designers leveraged the HFLCC "fail off" design architecture to provide a safe "revert to" configuration by giving the pilot direct control of the BYPASS switches on the motherboards. When the pilot selected the BYPASS mode of operation, the HFLCC bypassed the digital modules on all four branches, resulting in a configuration that was functionally equivalent to that of the unmodified FLCC. The pilot controlled the BYPASS mode via an up-front control (UFC) entry on the DED.

HFLCC Flight Test Program

The HFLCC test program was broken into two phases of flight test. The first consisted of 13 test flights from April to May of 2015. It demonstrated promising results for both the flying qualities of the HFLCC and the ability of the system to host digital applications. Operation of the HFLCC throughout the F-16 flight envelope demonstrated no degradation in F-16 flying qualities or handling qualities while the HFLCC was fully functioning in the PASSTHROUGH mode. Additionally, the initial flight test of select Auto GCAS and PARS maneuvers in the first phase showed that not only was HFLCC able to seamlessly incorporate a digital application, but the Auto GCAS performance was equivalent to the already-fielded digital Auto GCAS on Block 40/50 F-16 aircraft.

A second phase of HFLCC test was conducted from January to March of 2016, consisting of 16 test flights. The primary purpose of the second phase was to fully evaluate the performance of Auto GCAS on the HFLCC and evaluate minor deficiencies with the HFLCC that were addressed by the second software release. Additionally, improvement to Auto GCAS functionality during split-S maneuvers, as described earlier in this paper, was incorporated and tested during the second test phase. Although the overall system was considered mature and performed as expected during the majority of the second phase of flight test, safety planning was implemented to prevent reliance on the system under test. Two instances of correctly exercised safety mitigations are discussed in this paper. On one of the test flights, the primary method for the pilot to override the automatic system malfunctioned and required safety mitigations to be exercised. Additionally, on a different test flight, during the first attempt of the fastest, steepest diving test point, pre-planned safety procedures proved critical in protecting the pilot when an unanticipated system failure occurred.

Paddle Switch Functionality

During flight test of both the digital Auto GCAS and the HFLCC it was critical for the pilot to have the capability to take immediate control of the aircraft and the system under test. The paddle switch was mechanized for this purpose. When depressed, it would terminate any ongoing automatic maneuver and give the pilot immediate control. The paddle switch, shown in figure 12, was located on the control stick and exercised at the beginning of every test flight. While paddle switch functionality was evaluated on the first system checkout flight of both phases of HFLCC, on subsequent test flights the paddle switch check was treated as a quick hardware confirmation. This verification was pilot paced and involved minimal control room input or data review.

Figure 12 Paddle Switch to Disengage Auto GCAS

On the twenty-fourth flight of the HFLCC test program, the test pilot initiated the standard paddle switch verification check. This involved slightly banking the aircraft, activating a PARS maneuver, and quickly disengaging the maneuver with the paddle switch. Though the paddle switch had appeared to correctly disengage PARS/Auto GCAS on every previous test missions, on this mission the aircraft appeared to continue rolling past wings-level despite the pilot engaging the paddle switch. The pilot quickly commented on the aircraft dynamics, noted that it may have been due to the fact that this was the first time the program was testing clean loading configuration on the aircraft, and then proceeded to quickly attempt a second paddle switch check. The second paddle switch engagement appeared to work correctly, and the pilot proceeded to wait for guidance from the control room.

It had not been standard for the control room to do a detailed real-time evaluation of paddle switch functionality. Due to the pilot's comment on the unexpected aircraft dynamics during the first paddle switch check, the control room directed the pilot to pause and wait for further data review prior to proceeding to the first test card. This review uncovered that while the paddle switch was correctly depressed during the automatic maneuver and a message was sent to the HFLCC to terminate the maneuver, the flight controls continued to command a roll rate until the pilot countered with lateral stick input. This was treated as an unexpected event. Further, paddle switch functionality was a critical safety feature that allowed the pilot to take control away from the new flight control system under test. The team terminated testing, had the pilot put the system into the BYPASS mode and returned to base.

Following this unexpected test event, the team stood down testing and not only reviewed the data and system functionality to determine the root cause of the anomaly, but also reevaluated the way the paddle switch check was treated at the beginning of the flight. Review of the data and the system design determined that the cause of the paddle switch anomaly was due to the PARS being activated at nearly 0 degrees flight path angle, a boundary condition that caused the system to consider both nose-high and nose-low PARS activations. The system subsequently continued to command a roll rate even though the pitch rate and maneuver command were correctly terminated. Fortunately, detailed review of the system design discovered that this scenario occurred only near 0 degrees flight path angle (FPA), which was not a specific test condition for any remaining test. Additionally, the worst-case scenario for this paddle switch anomaly was determined to be acceptable for safely continuing the remainder of testing. Bank-angle limits existed within the software, and the pilot was readily able to terminate the roll rate with lateral stick input.

More important, the team reevaluated the way that flight test of the safety critical features, including the paddle switch, were conducted at the beginning of each flight. The paddle switch functionality was being treated as a hardware check, with the assumption that once the system check flight had been performed, all that would be required on subsequent flights would be to ensure that the electronic connection from the paddle switch to the HFLCC was working. However, the paddle switch functionality was both hardware- and software-dependent since the response changed depending on aircraft state. The team realized that each check should have been treated as a test point with detailed data review. Subsequent review of data from previous HFLCC test missions determined that the same paddle switch anomaly had occurred several times before but had not been noticed by the pilot or the engineers.

Several lessons were learned from the paddle switch anomaly on HFLCC and have been implemented on other test programs. Nothing is ever 100 percent proven, and despite previous positive performance the test team should not be lured into a false sense of security. Additionally, system safety checks should be treated as dedicated test points with detailed review. Finally, a test team should always understand system boundary conditions and design test points to explore those conditions, and should be prepared for anomalous behavior.

Auto GCAS: Safely Testing Mishap Recreation

The bulk of the flight test conducted on Auto GCAS, both on the fielded digital system and the HFLCC, was to test different aircraft diving maneuvers and ensure that Auto GCAS would correctly activate and provide ground clearance. A variety of aircraft states, airspeeds, and maneuvers were tested, to include a large number of historical mishaps. These mishaps were recreated to evaluate whether Auto GCAS would prevent collision with the ground. At no time did the team ever rely on Auto GCAS-hosted as the system under test, to provide safe ground clearance. To ensure sufficient pilot reaction time in the event of adverse system response, the terrain used by Auto GCAS was raised by a TCB, as described previously in this paper. Additionally, abort conditions were used in order to ensure that the pilot had at least four seconds to react and recover the aircraft if Auto GCAS did not activate. This was especially critical for test conditions with steep dive angles and high descent rates.

For the second phase of HFLCC the team focused on Auto GCAS performance evaluation. The majority of the test cards, especially the more challenging ones, had been flown numerous times on previous test programs. Additionally, test pilots had rehearsed the test conditions in simulation facilities prior to flight test. Finally, for all of the Auto GCAS test cards, active control room monitoring of aircraft limits and abort conditions was required. Together, all these factors resulted in the test program being categorized as low-risk by the safety review board.

One of the most challenging Auto GCAS test points was a mishap recreation of a historical G-LOC scenario. The test condition involved achieving 630 knots calibrated airspeed (KCAS) with an 85 degree dive angle. In order to retain sufficient pilot reaction time should Auto GCAS operate incorrectly, an activation was expected at 33,000 feet and the terrain was artificially raised by 6,600 feet with a TCB. Additionally, the control room monitored multiple abort parameters, including an altitude abort at 26,000 feet, and provided a "watch airspeed call" at 700 KCAS to ensure the aircraft would not exceed the 800 KCAS operational limits. This test point, with all the parameters described above, is shown in figure 13.

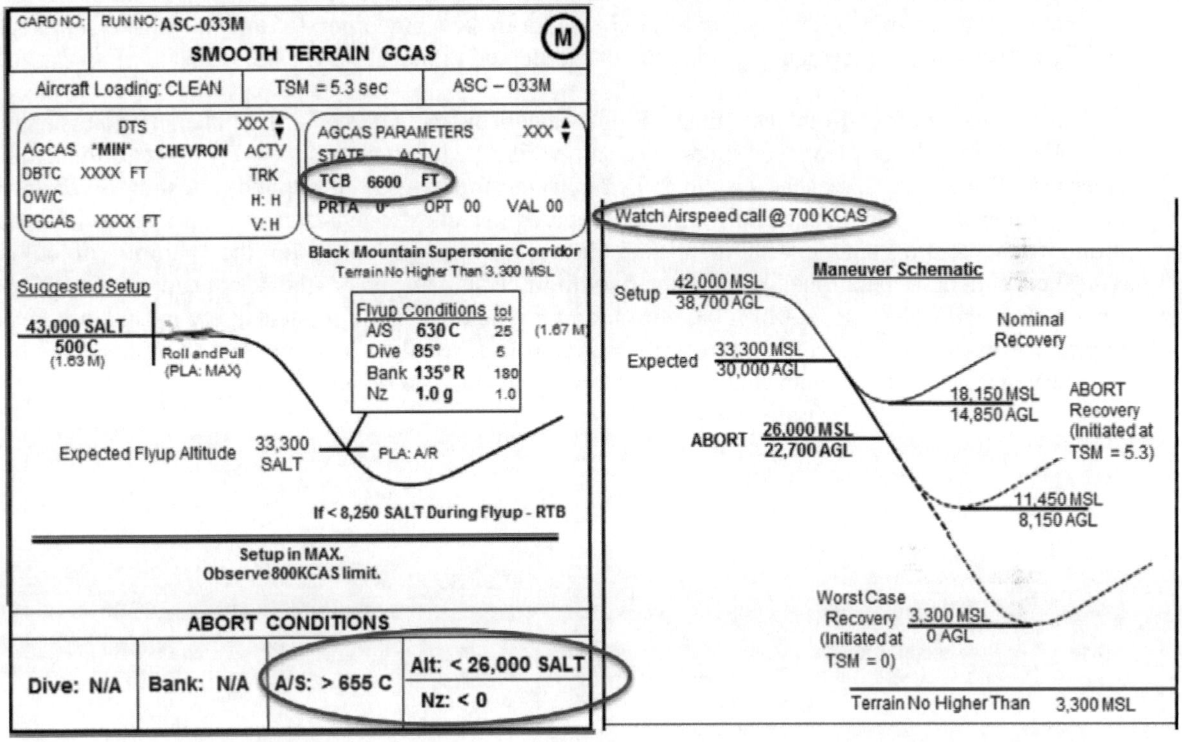

Figure 13 Auto GCAS Test Card Recreating Historical GLOC Mishap

The test point shown in figure 13 represents the fastest, steepest diving Auto GCAS test point. The pre-planned safety procedures proved critical in protecting the pilot. An Auto GCAS failure occurred prior to the activation expected at 33,000 feet. The control room quickly made an abort call, and the pilot immediately started to recover the aircraft. During recovery the pilot noted that the system had failed, through indications in the head-up display. Since Auto GCAS was never relied upon for safety and appropriate safety planning was put in place, the team was able to continue testing despite adverse system performance. A reiteration of the test point resulted in nominal Auto GCAS performance. Subsequent analysis determined that the steep dive with a high-g entry was one of the most challenging scenarios for Auto GCAS; it was very dynamic with many parameters changing quickly. This was the most strenuous situation for Auto GCAS algorithm processing. Currently HFLCC Auto GCAS is being improved in this scenario and will be tested prior to fielding.

CONCLUSIONS AND FUTURE OF HFLCC

The HFLCC test program successfully completed testing in March 2016. Overall, the performance of HFLCC and the ability to host Auto GCAS on pre-Block 40 aircraft was encouraging. Both of the anomalies discovered during flight test (the paddle switch anomaly and Auto GCAS failure) were understood by the end of the program and are currently being fixed for the future fielded system. The current goal is for Auto GCAS capabilities to be hosted on all U.S. Air National Guard pre-Block 40 aircraft by 2020. A subsequent flight test program on Block 30 F-16s with the HFLCC will start in late 2017. Additionally, incorporation of HFLCC on the Block 30 aircraft provides the capability to host other digital applications. Finally, improvements to the split-S functionality of Auto GCAS are being incorporated into the HFLCC application on Block 30 aircraft and are being considered for the fielded digital Auto GCAS.

Lessons learned from flight test of the HFLCC have been applied to other flight test programs ongoing at the 416 FLTS. The test team continues to remain vigilant when it comes to safety and trusting a system under test. The motto, "Treat the system as if it is trying to kill you," is applied to testing. Even for test programs deemed low-risk, the team remains prepared for unexpected aircraft results, especially when testing flight controls and executing mishap recreations. Additionally, for the majority of collision avoidance and non-operationally representative flight test maneuvers, the test team continues to be conservative and utilizes a control room with monitored abort conditions as a safety mitigation. Finally, despite previous positive performance of safety critical features, the team continues to treat safety checks as dedicated test conditions with control room monitoring and data review.

ACKNOWLEDGEMENTS

The authors would like to thank every member of the Air Force Test Center and Global Power Fighters Combined Test Force who played a role in the success of the F-16 Auto GCAS test programs, as well as the many hundreds of additional people who have contributed to making the F-16 Auto GCAS a reality over the course of several decades. A special thanks to test team members Maj. Corey "Rufus" Florendo, Andrew Whitten, Jackie Pajak, Collin Drake, Mike Perron, Rebecca Mitchell, Lt. Timothy Boersig, P. Travis Millet, Christian Garcia, Marla McCully, Shawn Whitcomb, David Stoner, Russell Turner, Ed Griffin, Finley Barfield, Amy Burns, Don Swihart, and Mark Skoog. Finally, the authors would also like to dedicate this paper to all the F-16 aircrew that have lost their lives in controlled flight into terrain or spatial-disorientation mishaps:

Captain William Dubois, 2014
Lt Col (Ret) Matthew LaCourse, 2014
Captain James Steel, 2013
Captain Lucas Gruenther, 2013
Captain Eric Ziegler, 2011
Captain George Houghton, 2009
Second Lieutenant David Mitchell, 2008
Lieutenant Colonel Kevin Sonnenberg, 2007
Major Troy Gilbert, 2006
Lieutenant Colonel Dillon McFarland, 2002
Captain Benton Zettel, 2002
Captain Luke Johnson, 2002
Major Aaron George, 2001
Mr. Judson Brohmer, 2001
Captain Mitchell Bulmann, 2001
First Lieutenant Randolph Murff, 2001
Major Stephen Simmons, 2000
Major Brison Phillips, 2000
Second Lieutenant Patrick Potter, 1998
Captain Keith Sands, 1998
Major Pete Woodbury, 1997
Captain Mark McCarthy, 1995
Second Lieutenant Julian Beneker, 1994
First Lieutenant Stephen Taylor, 1993
Captain John Barelka, 1993
Captain James Reynolds Junior, 1993
Mr. Joe Bill Dryden, 1993
Captain Luis Jordan, 1993
Captain Glen Porter, 1993
Major Robert Verdone, 1993
Captain Donald Leckrone Junior, 1992
Lieutenant Colonel Edward Hackney, 1992
Lieutenant Colonel Jeff Dennis, 1991
Staff Sergeant Chris Ford, 1991
Captain Arnold Clarke, 1991
Captain Michael Sowell, 1991
Captain Percival Gates III, 1991
Captain Keith Nylander, 1991
Major Robert Ashenfelter, 1991
Captain Michael Chinburg, 1991
Lieutenant Colonel Barry Bost, 1990
Major Archie Stuart, 1990

And at least a dozen more in the 1980s.

FULL ENVELOPE AIRSPEED CALIBRATION USING A SINGLE TURNING ACCELERATION TECHNQIUE

Juan D. Jurado[a,*], Nick W. Beihl[b,**], Clark C. McGehee[c]

[a]*Air Force Institute of Technology, Wright-Patterson AFB, OH*
[b]*5th Special Operations Squadron, Hurlburt Field, FL*
[c]*US Air Force Test Pilot School, Edwards AFB, CA*

Abstract

Virtually every performance or flying qualities test program requires the calibration of either a production air data system or a flight test Air Data Boom (ADB). Classical Pitot-static calibration techniques include the Ground Speed Course and Cloverleaf. However, such methods are highly sensitive to maneuver execution and data collection errors. Additionally, they are more time consuming due to their simplified data reduction algorithms, which require multiple test points across a range of airspeeds and aircraft configurations in order to model the speed dependencies in Pitot-static system errors. This paper proposes a Turning Acceleration Flight Test Technique (FTT) that can yield the airspeed calibration results for the entire flight envelope using a single accelerating turn. Leveraging state-of-the-art data collection equipment and statistical parameter estimation methods, the proposed FTT allows for the computation of airspeed correction coefficients as functions of airspeed using a single test maneuver. The resulting computations also yield confidence intervals on the coefficient estimates, giving users critical feedback on the quality of test execution and data collection. Additionally, due to the advanced mathematical algorithms provided, variations in airspeed and heading that would normally be considered sources of error become valid sources of additional data. The proposed technique is shown to reduce flight test time by 87% when compared to the Ground Speed Course and 57% when compared to the Cloverleaf using real-world flight test data from a C-130J Flying Qualities test program. As an update to previous publications, additional lessons learned and guidance are provided in the area of sample rate selection through the analysis of autocorrelation of flight test data. If widely adopted, the Turning Acceleration FTT has the potential to save millions of dollars for testers worldwide.

*Corresponding author.
**Member, SETP
Email addresses: `juan.jurado.1@us.af.mil` (Juan D. Jurado), `nbeihl@hotmail.com` (Nick W. Beihl), `clark.mcgehee.1@us.af.mil` (Clark C. McGehee)

96TW-2015-0276: Approved for public release; distribution is unlimited
60th SETP International Symposium, 21-24 September 2016, Anaheim, California

CONTENTS

LIST OF SYMBOLS

M . . . . Number of measurements

N . . . . Number of test points

R . . . . Model residual (error)

S . . . . Sample size

V_{CAS} . . . . Calibrated airspeed

$V_{\mathrm{G_c}}$ . . . . Computed groundspeed

$V_{\mathrm{G_m}}$ . . . . Measured groundspeed

V_{G} . . . . Groundspeed

V_{IAS} . . . . Indicated airspeed

V_{TAS} . . . . True airspeed

V_{W} . . . . Wind velocity

ΔV . . . . Airspeed correction

$\bar{X}$ . . . . Average of X

$\mathbf{y}$ . . . . Model outputs

$\mathbf{\Phi}$ . . . . Model feature matrix

$\boldsymbol{w}$ . . . . Model weights

δ_k . . . . Model relative AICc

$\mathcal{L}$ . . . . Model maximized likelihood function

$\mathcal{M}_{\Delta}$ . . . . Airspeed correction model

ϕ . . . . Ground track

π . . . . Akaike weight

ψ . . . . Aircraft heading

σ . . . . Standard deviation

φ_{W} . . . . Wind direction

k . . . . Number of models

p . . . . Number of model parameters

LIST OF ACRONYMS

ADB	Air Data Boom
AIC	Akaike Information Criterion
AICc	Corrected Akaike Information Criterion
AOA	Angle of Attack
AOB	Angle of Bank
CAS	Calibrated Airspeed
CI	Confidence interval
E	East
FTT	Flight Test Technique
GPS	Global Positioning System
GS	Ground Speed
IAS	Indicated Airspeed
KCAS	Knots Calibrated Airspeed
KIAS	Knots Indicated Airspeed
N	North
NTPS	National Test Pilot School
RSSE	Root sum squared error
SSPE	Static Source Position Error
TAS	True Airspeed

1. INTRODUCTION AND BACKGROUND

Virtually every performance or flying qualities flight test program begins with the calibration of either the aircraft's production Pitot-static system and/or a dedicated ADB. One of the main goals in Pitot-static calibration is to determine the Static Source Position Error (SSPE), or the error induced in Pitot-static instrument readings due to the disturbance of the local static pressure field by the aircraft itself [1]. SSPE is typically manifested in errors in Indicated Airspeed (IAS) and altitude, both of which rely on accurate static pressure readings. Although there are currently many FTTs designed to collect Pitot-static calibration data, they are all based on the general principle of flying the aircraft at known conditions such as Ground Speed (GS) and altitude, while simultaneously comparing the known conditions to instrument readings.

Altitude calibration techniques pose significantly less challenges due to the ease and simplicity associated with flying the aircraft at a known geometric and pressure altitude using the prolific Tower Flyby FTT. In contrast, airspeed calibration techniques present additional challenges due to the common inability to measure (and differentiate between) IAS, GS, and wind speed during test execution. Additionally, since the static field disturbances that generate SSPE are produced by the act of flying the aircraft, the airspeed and altitude errors found during calibration tend to be functions of airspeed itself [1]. Therefore, in order to characterize SSPE in both airspeed and altitude, a range of test points covering a reasonable portion of the aircraft's flying envelope must be flown.

This paper aims to provide a robust framework for the data collection, reduction, and processing of airspeed calibration data as well as a novel FTT for obtaining such data. The combination of the proposed data framework and method for obtaining data will be shown to drastically improve the test efficiency and data quality associated with the airspeed calibration of any flight test program. Prior to identifying the underlying gaps in the state of the art methods, it is important to briefly compare the existing methods.

1.1 Current Methods

As previously mentioned, most current FTTs focus on flying the aircraft at known conditions and comparing the known parameters to instrument readings. Additionally, airspeed calibration poses a more challenging problem when compared to altitude calibration due to the inability to directly measure Calibrated Airspeed (CAS), especially in the presence of wind. Up to date, the challenge of measuring CAS has been solved by either flying at or near Sea Level (which allows for the approximation of CAS using True Airspeed (TAS) or GS and wind), or calculating CAS using TAS, altitude, and indicated temperature. As such, the following sections focus on how some of the current FTTs solve the challenge of measuring CAS.

1.1.1 Ground Speed Course

In [1], Erb describes the classical Ground Speed Course, which prescribes flying a two-leg, reciprocal heading, straight line course that maintains airspeed and heading while allowing the aircraft ground track to drift with wind. The purpose of the reciprocal headings is to "average out" wind speed using the assumption that wind speed and direction remain constant during test execution. This allows for the approximation of CAS given an average computed GS. The technique is then repeated for a series of IASs within the aircraft's flying envelope.

1.1.2 Three-legged GPS Course

In [2], National Test Pilot School (NTPS) improves on the Ground Speed Course by taking advantage of Global Positioning System (GPS) as a truth source for GS and Ground Track. The Three-Legged GPS FTT prescribes three orthogonal straight-line legs where airspeed and heading and tightly held. Once flown, basic trigonometric formulas are used to determine wind speed and direction, which allow the computation of TAS and CAS. Again, this technique is repeated for every airspeed of interest in the aircraft's flying envelope.

1.1.3 Cloverleaf and Level Turn

A subsequent and logical improvement to the previously listed FTTs involves taking advantage of the high-speed data recording capabilities now found in most flight test programs, as well as linear regression or statistical computation methods. As such, [3] and [4] provide the Cloverleaf and Level Turn FTTs, respectively. Both FTTs prescribe flight paths where the aircraft's heading is varied to provide a linear regression matrix with adequate measurement observability to estimate CAS, wind speed, and wind direction. These FTTs not only allow for but prescribe variations in heading during test execution, which makes test execution less difficult. However, IAS must still be tightly held since the techniques must be repeated for each desired airspeed in the aircraft's flying envelope. Finally, the use of regression algorithms expands the possibilities for data quality feedback by allowing the computation of confidence intervals on the results.

1.2 Problem Formulation

The calibration techniques described above all aim at producing CAS and wind speed measurements that can be used to determine the SSPE for a particular airspeed when compared to the aircraft's instrument indications. The newest techniques (Cloverleaf and Level Turn) leverage modern data collection systems, relax the necessary tolerances on heading, and utilize statistical regression techniques to produce least-squares solutions and confidence intervals. However, airspeed must still be tightly held and the technique must be flown for each airspeed of interest within the aircraft's flying envelope.

1.3 Research Contributions

The Cloverleaf and Level Turn FTTs are able to collect the necessary data to differentiate TAS (and subsequently CAS) from wind speed by providing the regression matrix with measurement observability in heading. That is, by changing heading during the execution of the test point, the regression matrix is able to accurately decompose the measured CAS into actual CAS and wind speed components.

The research provided in this paper expands the concept of measurement observability into the airspeed domain. Using the Level Turn FTT and associated regression matrices described by Jorris in [4], a novel FTT referred to from hereupon as the Turning Acceleration FTT, can be developed. In the Turning Acceleration FTT, heading and airspeed are uniformly changed throughout the test point by accelerating the aircraft while executing the turn. The result is an airspeed calibration technique that not only tolerates, but prescribes changes in heading and airspeed throughout the test point. More importantly, the new technique is able to produce the entire SSPE curve for an aircraft using only a single test point.

The accompanying regression algorithm explained further in later sections utilizes Akaike Information Criterion (AIC) as described in [5] and [6] to decide on the most appropriate SSPE curve shape for each aircraft from one of many pre-programmed family of shapes. Additionally, it provides a confidence interval around the curve itself, which can be used to determine data quality and algorithm confidence. Finally, a Monte-Carlo experiment focusing on the effects of wind variability is provided for three example aircraft to give the user an accurate tradeoff analysis between wind variability and expected confidence interval size.

1.4 Outline

Section 2 defines the data requirements, provides a generalized data reduction process for airspeed calibration, and discusses the merits of several widely-used techniques before proposing the Turning Acceleration technique that is the focus of this paper. Section 3 describes practical considerations involved in collecting the data required for the methods described in 2. Section 4 provides and analyzes numerical results using high-fidelity simulated data of three different example aircraft as well as real-world data from a C-130J flight test program. Finally, Section 5 summarizes the research team's conclusions and suggestions for future work on the subject area.

2. METHODOLOGY

The problem that all airspeed calibration methods seek to solve is to define SSPE (ΔV), which is simply the error between IAS (V_{IAS}) and CAS (V_{CAS}) in the airspeed case. ΔV can then be used to correct V_{IAS} and provide the pilot with an accurate estimate of V_{CAS} using the relationship

$$V_{\text{CAS}} = V_{\text{IAS}} + \Delta V. \tag{1}$$

In general, ΔV varies with V_{CAS}. As V_{CAS} itself is not measurable, the challenge in airspeed calibration is to design an experiment in which V_{CAS} can be inferred from other measurable data. Observations of ΔV over a range of V_{CAS} are then used to build an approximate model of ΔV as a function of either V_{CAS} or V_{IAS}. This model then becomes the source of ΔV in practical applications of (1).

To maintain focus on these practical applications, the theoretical background presented in this section will be tied to pseudocode modules, offering the reader a skeleton

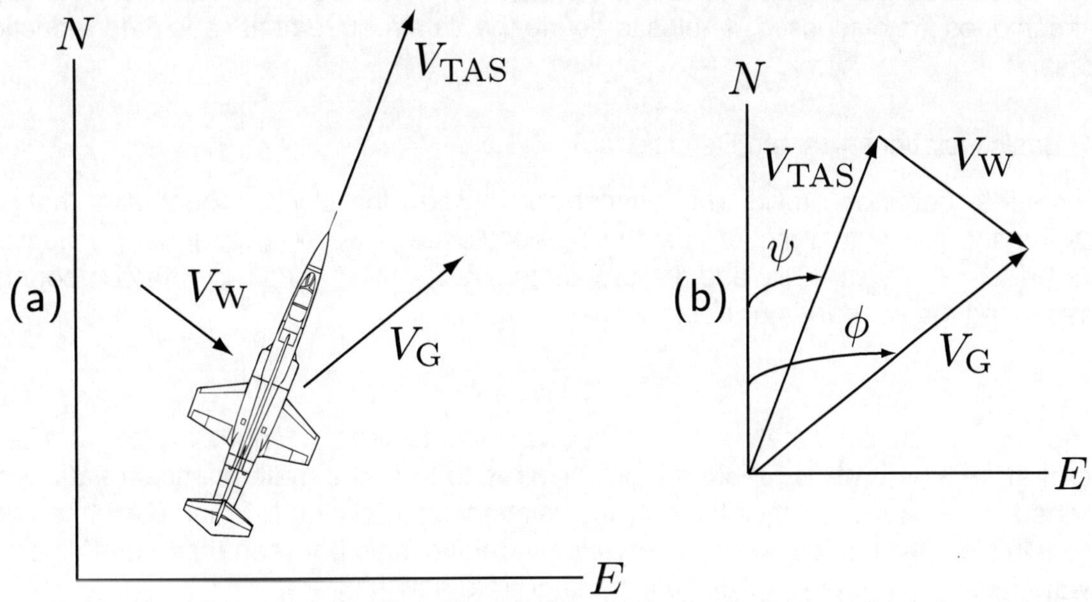

Figure 1: Definitions of TAS (V_{TAS}), wind speed (V_{W}), and GS (V_{G}) vectors. In (a), a cartoon of an F-104 Starfighter is shown to present the windspeed, TAS, and GS vectors intuitively. In (b), the vectors are exaggerated to highlight their mathematical relationship, and to define the aircraft's heading (ψ) and course along the ground (ϕ). Not shown is the wind angle with the north axis (φ_{W}).

for implementing the methods described by this paper. On an aircraft, equation (1) could be programmed into a mission computer to provide an estimate of CAS to the pilot as detailed in Program 1.

Program 1 Mission Computer Airspeed Correction

```
procedure CORRECTAIRSPEED(V_IAS)
    ΔV ← AIRSPEEDERRORFUNCTION(V_IAS)        ▷ Not yet defined
    V_CAS ← V_IAS + ΔV
    return V_CAS
end procedure
```

The next task is to define the airspeed error function.

2.1 Computing the SSPE Curve

The focus of this subsection is to develop the machinery required to compute a SSPE curve from a very generalized set of test data, assuming that these test data are available. It provides a common framework within which various techniques for collecting the flight test data may be compared on equal terms and allows for the computation of SSPE curves with fewer assumptions than traditional data reduction techniques. The mechanics of actually collecting the flight test data are discussed in subsection 3, but before the flight test techniques are discussed, it helps to define the data requirements and data reduction process.

2.1.1 Basic Relationships and Definitions

It is useful to begin by providing relationships between the physical quantities that describe the aircraft's planar motion. Let the TAS vector be $V_{\text{TAS}} = (V_{\text{TAS}_\text{E}}, V_{\text{TAS}_\text{N}})^\top$, the wind vector be $V_\text{W} = (V_{\text{W}_\text{E}}, V_{\text{W}_\text{N}})^\top$, and the GS vector be $V_\text{G} = (V_{\text{G}_\text{E}}, V_{\text{G}_\text{N}})^\top$, with components defined along the east and north axes. Then

$$V_{\text{TAS}} + V_\text{W} = V_\text{G} \tag{2}$$

describes the relationship between the three vectors, as seen in Figure 1. Note that TAS, rather than IAS or CAS, is required if one desires to form the mathematical relationship described in (2). However, thus far, only the relationship between IAS and CAS has been defined, in (1). One can construct a very simple relationship between these quantities as follows. First, from [1], the explicit definitions of CAS and TAS are

$$V_{\text{CAS}} = a_0\sqrt{5\left[\left(\frac{q_c}{P_0}+1\right)^{\frac{2}{7}}-1\right]} \tag{3}$$

$$V_{\text{TAS}} = a_0\sqrt{5\frac{T}{T_0}\left[\left(\frac{q_c}{P}+1\right)^{\frac{2}{7}}-1\right]} \tag{4}$$

where a_0 is the speed of sound at sea level on a standard day, T is the static air temperature, T_0 is the temperature at sea level on a standard day, q_c is the impact pressure, P_0 is the pressure at sea level on a standard day, and P is the static pressure. Note that when $T = T_0$ and $P = P_0$, $V_{\text{CAS}} = V_{\text{TAS}}$. Then, by restricting oneself to collecting data at or near ambient conditions (P_0, T_0), one may simply interchange CAS and TAS. Relaxing this assumption is certainly possible, but it requires the use of (4) for the remainder of the discussion. For simplicity, it is assumed that the test is conducted at (P_0, T_0) for the remainder of this discussion.

Given the $V_{\text{CAS}} = V_{\text{TAS}}$ assumption, and a definition of the IAS vector as $V_{\text{IAS}} = (V_{\text{IAS}_\text{E}}, V_{\text{IAS}_\text{N}})^\top$, substituting (1) into (2) yields an expression relating indicated airspeed to ground speed,

$$V_{\text{IAS}} + \Delta V + V_\text{W} = V_\text{G}. \tag{5}$$

In practice, it is difficult to measure the components of IAS and GS; instead, magnitude and direction are typically measured. The direction of both the TAS and IAS vectors relative to the north axis is the aircraft heading ψ, while the direction of the GS vector relative to the north axis is the aircraft course ϕ. Also, in general, the wind vector V_W is not measurable.

Given these limitations, and test measurements for V_{IAS} and V_G, the objective is to build a model $\mathcal{M}_\Delta$ of ΔV as a function of V_{IAS} so the aircraft's IAS may be corrected to

Table 1: Example model families used in fitting SSPE curves.

Model Name	Mathematical Form	Type
Polynomial(N)	$\mathcal{M}_\Delta(V_{\text{IAS}}; \boldsymbol{w}) = \sum_{i=0}^{P} w_i V_{\text{IAS}}^i$	Linear
Power	$\mathcal{M}_\Delta(V_{\text{IAS}}; \boldsymbol{w}) = w_0 + w_1 V_{\text{IAS}}^{w_2}$	Nonlinear
Exponential	$\mathcal{M}_\Delta(V_{\text{IAS}}; \boldsymbol{w}) = w_0 + w_1 \exp(w_2 V_{\text{IAS}})$	Nonlinear

CAS as shown in Program 1. Typically, $\mathcal{M}_\Delta$ will be parameterized by a set of P weights $\boldsymbol{w} = (w_1, \ldots, w_p)^\top$, which are inferred from the test data. A basic implementation is shown in Program 2, and some examples of suitable $\mathcal{M}_\Delta$ are shown in Table 1.

Program 2 Airspeed Error Function

```
data w                                      ▷ Computed before and cached.
data {M_Δ}                                  ▷ A handle to the SSPE model function.
procedure AIRSPEEDERRORFUNCTION(V_IAS)
    ΔV ← M_Δ(V_IAS; w)                      ▷ An arbitrary function of V_IAS and w.
    return ΔV
end procedure
```

2.1.2 Fitting the SSPE Model Curve

Like Program 1, Program 2 is quite simple to implement given $\boldsymbol{w}$ and $\mathcal{M}_\Delta$. The difficult task is obtaining these quantities. To begin, note that

$$\Delta V = \mathcal{M}_\Delta(V_{\text{IAS}}; \boldsymbol{w}) + R(V_{\text{IAS}}), \tag{6}$$

where $R(V_{\text{IAS}})$ is the residual (model error) in $\mathcal{M}_\Delta$ at V_{IAS}. Equation (6) can be rearranged as

$$R = \mathcal{M}_\Delta(V_{\text{IAS}}; \boldsymbol{w}) - \Delta V, \tag{7}$$

which can then be minimized in a least-squares sense over a set of N test data to compute an estimate of $\boldsymbol{w}$ as

$$\hat{\boldsymbol{w}} = \arg\min_{\boldsymbol{w}} \sum_{\forall n \in N} R_{(n)}^2 = \arg\min_{\boldsymbol{w}} \sum_{\forall n \in N} \left[\mathcal{M}_\Delta(V_{\text{IAS}_{(n)}}; \boldsymbol{w}) - \Delta V_{(n)}\right]^2. \tag{8}$$

Many algorithms are capable of solving (8); the MATLAB routine `fitnlm` was chosen for this particular application. Typically, such algorithms desire a set of known data $\boldsymbol{y}$, a feature matrix $\boldsymbol{\Phi}$ consisting of the model inputs where each column is a unique feature and each row is a unique observation, and a handle to a function that maps $\boldsymbol{\Phi}$ and a set of weights $\boldsymbol{w}$ to the outputs (typically written $(\boldsymbol{\Phi}, \boldsymbol{w}) \mapsto \mathcal{M}_\Delta(\boldsymbol{\Phi}; \boldsymbol{w})$). It would then use an iterative method, such as the Levenberg-Marquardt method, to converge on an estimate for $\hat{\boldsymbol{w}}$ that solves (8).

One can rearrange equation (5) to suit this purpose. Assuming S test data samples are available and taking V_{IAS}, V_{G}, ψ, and ϕ as knowns, ΔV and V_{W} as unknowns, and splitting the vectors into their north and east components, eq. (5) can be rearranged as

$$V_{\mathrm{G}} - V_{\mathrm{IAS}} = V_{\mathrm{W}} + \Delta V, \tag{9}$$

$$\begin{bmatrix} V_{\mathrm{G_N}} - V_{\mathrm{IAS_N}} \\ V_{\mathrm{G_E}} - V_{\mathrm{IAS_E}} \end{bmatrix} = \begin{bmatrix} V_{\mathrm{W_N}} + \Delta V_{\mathrm{N}} \\ V_{\mathrm{W_E}} + \Delta V_{\mathrm{E}} \end{bmatrix}, \tag{10}$$

$$\underbrace{\begin{bmatrix} V_{\mathrm{G}_{(1)}} \cos\phi_{(1)} - V_{\mathrm{IAS}_{(1)}} \cos\psi_{(1)} \\ \vdots \\ V_{\mathrm{G}_{(S)}} \cos\phi_{(S)} - V_{\mathrm{IAS}_{(S)}} \cos\psi_{(S)} \\ V_{\mathrm{G}_{(1)}} \sin\phi_{(1)} - V_{\mathrm{IAS}_{(1)}} \sin\psi_{(1)} \\ \vdots \\ V_{\mathrm{G}_{(S)}} \sin\phi_{(S)} - V_{\mathrm{IAS}_{(S)}} \sin\psi_{(S)} \end{bmatrix}}_{\stackrel{\mathrm{def}}{=} \boldsymbol{y} \in \mathbb{R}^{2S}} = \begin{bmatrix} V_{\mathrm{W_N}} \\ \vdots \\ V_{\mathrm{W_N}} \\ V_{\mathrm{W_E}} \\ \vdots \\ V_{\mathrm{W_E}} \end{bmatrix} + \underbrace{\begin{bmatrix} \Delta V_{(1)} \cos\psi_{(1)} \\ \vdots \\ \Delta V_{(S)} \cos\psi_{(S)} \\ \Delta V_{(1)} \sin\psi_{(1)} \\ \vdots \\ \Delta V_{(S)} \sin\psi_{(S)} \end{bmatrix}}_{\approx \mathcal{M}_{\Delta}(\boldsymbol{\Phi}; \boldsymbol{w}) \in \mathbb{R}^{2S}}, \tag{11}$$

$$\boldsymbol{y} \approx \begin{bmatrix} V_{\mathrm{W_N}} \\ \vdots \\ V_{\mathrm{W_N}} \\ V_{\mathrm{W_E}} \\ \vdots \\ V_{\mathrm{W_E}} \end{bmatrix} + \mathcal{M}_{\Delta}(\boldsymbol{\Phi}; \boldsymbol{w}). \tag{12}$$

Equation (12) holds for all reasonable choices of $\mathcal{M}_{\Delta}$, including those shown in Table 1. Note the definitions of $\boldsymbol{y}$ and $\mathcal{M}_{\Delta}(\boldsymbol{\Phi}; \boldsymbol{w})$. Also, note that if wind effects are not negated by the flight test technique used to obtain the test data, values for the components of V_{W} will also have to be obtained when fitting $\hat{\boldsymbol{w}}$. This may be accomplished by simply replacing the right hand side of eq. (12) with

$$\boldsymbol{y} \approx \begin{bmatrix} 1 & 0 \\ \vdots & \vdots \\ 1 & 0 \\ 0 & 1 \\ \vdots & \vdots \\ 0 & 1 \end{bmatrix} \begin{bmatrix} V_{\mathrm{W_N}} \\ V_{\mathrm{W_E}} \end{bmatrix} + \mathcal{M}_{\Delta}(\boldsymbol{\Phi}; \boldsymbol{w}) \tag{13}$$

and taking $V_{\mathrm{W_N}}$ and $V_{\mathrm{W_E}}$ as free parameters when fitting the model. All that remains is to define $\boldsymbol{\Phi}$ before the reader is fully equipped to fit a model to test data. Unlike $\boldsymbol{y}$, $\boldsymbol{\Phi}$ depends on the particular model chosen. For example, the Polynomial(N) model has

$$
\mathbf{\Phi} = \text{diag}\left(\begin{bmatrix} \cos\psi_{(1)} \\ \vdots \\ \cos\psi_{(S)} \\ 0 \\ \vdots \\ 0 \end{bmatrix}\right)\begin{bmatrix} 1 & V^1_{\text{IAS}_{(i)}} & \cdots & V^P_{\text{IAS}_{(i)}} \\ \vdots & \vdots & & \vdots \\ 1 & V^1_{\text{IAS}_{(S)}} & \cdots & V^P_{\text{IAS}_{(S)}} \\ 0 & 0 & \cdots & 0 \\ \vdots & \vdots & & \vdots \\ 0 & 0 & \cdots & 0 \end{bmatrix}
+ \text{diag}\left(\begin{bmatrix} 0 \\ \vdots \\ 0 \\ \sin\psi_{(1)} \\ \vdots \\ \sin\psi_{(S)} \end{bmatrix}\right)\begin{bmatrix} 0 & 0 & \cdots & 0 \\ \vdots & \vdots & & \vdots \\ 0 & 0 & \cdots & 0 \\ 1 & V^1_{\text{IAS}_{(i)}} & \cdots & V^P_{\text{IAS}_{(i)}} \\ \vdots & \vdots & & \vdots \\ 1 & V^1_{\text{IAS}_{(S)}} & \cdots & V^P_{\text{IAS}_{(S)}} \end{bmatrix}. \tag{14}
$$

Since the polynomial model is linear, that is $\mathcal{M}_{\Delta}(\mathbf{\Phi}; \boldsymbol{w}) = \mathbf{\Phi}\boldsymbol{w}$, in the no-wind case, equation (13) can rewritten as

$$
\boldsymbol{y} \approx \mathbf{\Phi}\boldsymbol{w} \tag{15}
$$

and solved[1] in a least-squares sense as

$$
\boldsymbol{w} = (\mathbf{\Phi}^\top \mathbf{\Phi})^{-1} \mathbf{\Phi}^\top \boldsymbol{y}. \tag{16}
$$

In the nonlinear case, as for the exponential and power families of models, an iterative method is required, so for the sake of generality, the routine called "FITNLM" will be referenced when solving for $\boldsymbol{w}$. One could repeat this process and arrive at a set of weights for several model families. The question is then which model or models to choose to best represent the SSPE curve.

2.1.3 Choosing the Best Model

Having fit several models to the same data, the next task becomes choosing which model best describes the process that generated the data. Many techniques are well-suited for this task. The technique favored by the authors and employed in this investigation is to use the Akaike Information Criterion (AIC),

$$
AIC = 2p - 2\log\mathcal{L}, \tag{17}
$$

where p is the number of parameters (i.e. coefficients) in the model, and $\mathcal{L}$ is the maximized value of the model's likelihood function. In practice, AIC must be corrected for a

[1] Note that the system is overdetermined, so the task is to choose the "best" solution.

finite sample size, and this corrected quantity is referred to as Corrected Akaike Information Criterion (AICc). AICc is defined as

$$AICc = AIC + \frac{2p(p+1)}{S-p-1} \tag{18}$$

where S is the sample size.

AIC and AICc have their roots in information theory, and interested readers are directed to [5] for the theoretical background. Briefly, AIC and AICc are measures of how much information is lost in replacing the generating process (i.e. truth or reality) with a model. It balances how well the model fits the data, represented by the log likelihood $\log \mathcal{L}$, with the complexity of the model, represented by the number of parameters p. Clearly, if forced to choose a single model from a finite set of models fit to a data set, one should choose that model with the least information loss and with reasonable complexity. However, AIC and AICc provide an additional benefit, in that they can be used to combine models. When it is sensible to do so, mixing models allows for less information to be lost than choosing only one model. To determine what is "sensible", one must compute the relative Akaike weights π_k of the candidate models.

Let $\delta_k = AICc_k - \min\{AICc_1, \ldots, AICc_K\}$ be the relative AICc for each of the K models. Then the Akaike weight of each model is given by

$$\pi_k = \frac{\exp\left(-\frac{1}{2}\delta_k\right)}{\sum_{\kappa=1}^{K} \exp(-\frac{1}{2}\delta_\kappa)}, \tag{19}$$

and by comparing the π_k one can determine which model or models best describe the test data. Note that this formulation of π_k ensures that the model weights are normalized (sum to 1). Thus the combined model may be represented as a linear combination of the individual models,

$$\mathcal{M}_\Delta(V_{\text{IAS}}; \{\boldsymbol{w}_k, \pi_k\}_{\forall k \in K}) = \sum_{\kappa=1}^{K} \pi_\kappa \mathcal{M}_\Delta\left(V_{\text{IAS}}; \boldsymbol{w}_\kappa\right). \tag{20}$$

This is critical because it allows the models to be fit independently of one another and combined after the fact.

In practice, some of the π_k may be small, indicating that the corresponding model is not very informative. It is useful to discard the non-informative models and recompute the π_k excluding those models. Doing so is a simple matter of reusing equation (19) with the δ_k corresponding to the offending models excluded.

2.1.4 Implementing the Method

Having now thoroughly discussed the technical background of the model, it is important to return to our original goal of characterizing the SSPE curve for a particular aircraft. In Program 1, it was shown (very simply) how a mission computer might take SSPE information and convert the aircraft's IAS V_{IAS} into CAS V_{CAS}. Then, Program 2 illustrated how the black-box function "AIRSPEEDERRORFUNCTION" might be implemented. The next task was to select a model and fit that model to a set of data collected during flight test. The mathematics of fitting and selecting models were briefly explained in the previous two subsections; it now makes sense to provide a sketch of how one might implement these ideas to achieve practical results. Refer to Program 3 for such an example.

Program 3 Fitting the SSPE Model

data $V_{\text{IAS}}, V_{\text{G}}, \psi, \phi$ ▷ Test data.
data $K = \{1, 2, \ldots, \text{NumberofModels}\}$ ▷ The set of model indices.
data $\{\mathcal{M}_{\Delta_1}, \ldots, \mathcal{M}_{\Delta_k}\}_{\forall k \in K}$ ▷ Function handles to all candidate models.
procedure CREATEMODEL
    $\mathbf{y} \leftarrow$ CREATELHS($V_{\text{IAS}}, V_{\text{G}}, \psi, \phi$) ▷ Create the vector of known data.
    for all $k \in K$ **do**
        $\mathbf{\Phi} \leftarrow$ CREATEFEATUREMATRIX(V_{IAS}, `ref` $\mathcal{M}_{\Delta_k}$) ▷ Specific to each model.
        $\{\hat{\boldsymbol{w}}_k, AICc_k\} \leftarrow$ FITNLM($\mathbf{\Phi}$, $\mathbf{y}$, `ref` $\mathcal{M}_{\Delta_k}$)
    end for
    for all $k \in K$ **do** ▷ Compute the initial δ_k.
        $\delta_k \leftarrow AICc_k - \min\{AICc_1, \ldots, AICc_K\}$
    end for
    $K' \leftarrow K$ ▷ Duplicate the set K as K'.
    for all $k \in K$ **do** ▷ Compute the initial π_k.
        $\pi_k \leftarrow \exp\left(-\frac{1}{2}\delta_k\right) \div \sum_{\kappa \in K} \exp(-\frac{1}{2}\delta_\kappa)$
        if $\pi_k < 0.1$ **then**
            $K' \leftarrow K' \setminus k$ ▷ Remove k from the set K'.
        end if
    end for
    for all $k \in K'$ **do** ▷ Compute new δ_k over K'.
        $\delta_k \leftarrow AICc_k - \min\{AICc_1, \ldots, AICc_K\}$
    end for
    for all $k \in K'$ **do** ▷ Compute new π_k over K'.
        $\pi_k \leftarrow \exp\left(-\frac{1}{2}\delta_k\right) \div \sum_{\kappa \in K'} \exp(-\frac{1}{2}\delta_\kappa)$
    end for
    if $\max\{\pi_k\} > 0.7$ **then** ▷ Arbitrary threshold value.
        $k \leftarrow$ INDEX($\max\{\pi_k\}$)
        return $\{(V_{\text{IAS}}) \mapsto \mathcal{M}_{\Delta_k}(V_{\text{IAS}}; \hat{\boldsymbol{w}}_k)\}$ ▷ Return a function handle to one model.
    else
        return $\left\{(V_{\text{IAS}}) \mapsto \sum_{k \in K'} \left[\pi_k \cdot \mathcal{M}_{\Delta_k}(V_{\text{IAS}}; \hat{\boldsymbol{w}}_k)\right]\right\}$ ▷ Return a function handle to a sum of all valid models.
    end if
end procedure

At this point, the machinery required to compute the SSPE curve from test data and to program a mission computer with that SSPE curve to obtain airspeed corrections in flight is available. All that remains is to collect the flight test data required to fit the models. A myriad of choices are available, and in the following section, a few existing methods are compared.

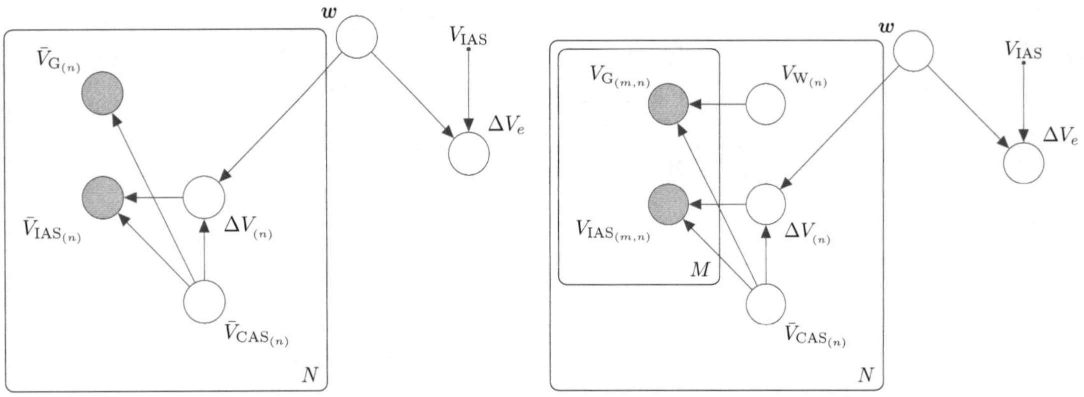

(a) Ground speed course method. (b) Level turn method.

Figure 2: Graphical representation of the model structure and dependencies of (a) the ground speed course and (b) the constant-speed turn method.

2.2 Classical Methods for Obtaining Flight Test Data

From the previous section, it is apparent that GS magnitude V_{G}, IAS magnitude V_{IAS}, heading ψ and ground track ϕ need to be collected during the flight test if one hopes to compute a SSPE curve. Also, wind effects must either be negated by the flight test technique, measured and incorporated into the model, or inferred from the test data. As mentioned before, it is difficult to measure the wind speed and direction reliably enough to arrive at passable results for the SSPE curve, so most techniques either focus on averaging out wind effects or ensuring that wind effects are observable enough to allow for inference of the wind speed and direction in post-processing. Also, if a high degree of accuracy is desired, then many useable samples must be taken during the flight test, either by completing a large number of test points or by leveraging the high sample rate of modern data acquisition equipment.

A brief discussion of three well-known techniques follows to motivate the technique proposed by this paper.

2.2.1 Ground Speed Course

The ground speed course method, as described in reference [1], uses a series of N test points, each conducted at a constant[2] IAS $\bar{V}_{\mathrm{IAS}_{(n)}}$, to form point estimates of ΔV at each $\bar{V}_{\mathrm{IAS}_{(n)}}$. Note that the $\bar{V}_{\mathrm{CAS}_{(n)}}$ cannot be measured directly; however, it can be inferred from the average GS $\bar{V}_{\mathrm{G}_{(n)}}$ and IAS $\bar{V}_{\mathrm{IAS}_{(n)}}$ from each run. Since wind effects are eliminated if the FTT is executed properly, from equation (2), $\bar{V}_{\mathrm{CAS}_{(n)}}$ is equal to $\bar{V}_{\mathrm{G}_{(n)}}$[3]. The weights $\boldsymbol{w}$ for $\mathcal{M}_{\Delta}(V_{\mathrm{IAS}};\boldsymbol{w})$ are then fit using the point estimates at the N airspeeds $\bar{V}_{\mathrm{IAS}_{(n)}}$ as training data.

In Figure 2(a), the dependencies of the ground speed course method are shown

[2]Assumed constant.

[3]This depends critically on the assumptions made in the previous subsection.

graphically. The observed variables $\bar{V}_{\mathrm{G}}$ and $\bar{V}_{\mathrm{IAS}}$ are shown as shaded nodes, while the latent variables V_{CAS}, ΔV, and $\boldsymbol{w}$ are shown as unshaded nodes. On the left side, the test data are contained within a plate indicating that there are N repeated observations corresponding to the N test points. Connected nodes are dependent on their parent node; for example, $\bar{V}_{\mathrm{IAS}}$ and $\bar{V}_{\mathrm{G}}$ are both dependent on $\bar{V}_{\mathrm{CAS}}$. Said another way, knowing $\bar{V}_{\mathrm{CAS}}$ allows $\bar{V}_{\mathrm{G}}$ and $\bar{V}_{\mathrm{IAS}}$ to be predicted. Conversely, if a child node is observed, the value of the parent nodes can be inferred. Thus, $\boldsymbol{w}$ is inferred from the N test data, and new values of ΔV may be predicted for arbitrary inputs V_{IAS}, as shown on the right side. Note that this graph does not offer details as to how inference or predictions are made, since its purpose is to highlight the structure of the model.

The primary drawbacks of the ground speed course method are that for the assumptions to hold, tight tolerances on airspeed and altitude must be held throughout the maneuver. Also, a single test point yields only a point estimate of ΔV at the prescribed V_{IAS}. As N is typically of order 10 due to practical constraints, parameter estimates $\hat{\boldsymbol{w}}$ formed typically have wide confidence intervals, yielding imprecise predictions of ΔV. Note that the data collected with this technique fit within the bounds of the data reduction framework proposed in the previous section. Concatenating the results of all N test points into a single vector and modifying equation (12), yields

$$\begin{bmatrix} V_{\mathrm{G}_{(1)}} - V_{\mathrm{IAS}_{(1)}} \\ \vdots \\ V_{\mathrm{G}_{(N)}} - V_{\mathrm{IAS}_{(N)}} \end{bmatrix} = \begin{bmatrix} \Delta V_{(1)} \\ \vdots \\ \Delta V_{(N)} \end{bmatrix} \approx \mathcal{M}_{\Delta}(\boldsymbol{\Phi}; \boldsymbol{w}), \tag{21}$$

where $\boldsymbol{w}$ can be fit for an arbitrary number of models $\mathcal{M}_{\Delta}$ as described above.

2.2.2 Level Turn and Cloverleaf Methods

The level turn method, as described in reference [4], and the cloverleaf method, as described in [3], are evolutions of the ground speed course method (and many other FTTs and are designed to leverage the high sample rates of modern instrumentation, data acquisition systems, and computational methods to obtain more precise predictions of ΔV. Both methods also use a series of N test points, each conducted at an average IAS $\bar{V}_{\mathrm{IAS}_{(n)}}$, but for each test point, M measurements of $V_{\mathrm{IAS}_{(m,n)}}$ and $V_{\mathrm{G}_{(m,n)}}$ are taken. Typically M is of order 10^3 or more; consequently, for each of the N test points, a probability distribution of $\Delta V_{(n)}$ is obtained rather than a simple point estimate. Additionally, wind effects may be incorporated into the model as an additional latent variable, so a probability distribution of $V_{W_{(n)}}$ is also obtained for each of the N test points; also, the models are tolerant of wind noise. The weights $\boldsymbol{w}$ for $\mathcal{M}_{\Delta}(V_{\mathrm{IAS}}; \boldsymbol{w})$ are then fit using all available data at the N airspeeds, yielding more precise parameter estimates than were afforded by the ground speed course method.

Figure 2(b) shows the dependencies of methods of this type. Clearly, they are similar to the ground speed course method. The primary structural differences are the addition of wind as a latent variable and the incorporation of many (M) realizations of the observed variables for each of the N test points. Thus, stronger estimates of the latent variables may be formed for the same number of test points N by using the level turn or cloverleaf methods rather than the ground speed course method. Note that the data collected with these and similar techniques also fit within the bounds of the data reduction framework proposed in the previous section. Again, taking ΔV to be constant for each test point, and modifying equation (12), yields

$$\begin{bmatrix} V_{\mathrm{G}_{(1)}} \cos\phi_{(1)} - V_{\mathrm{IAS}_{(1)}} \cos\psi_{(1)} \\ \vdots \\ V_{\mathrm{G}_{(M)}} \cos\phi_{(M)} - V_{\mathrm{IAS}_{(M)}} \cos\psi_{(M)} \\ \vdots \\ V_{\mathrm{G}_{(1)}} \sin\phi_{(1)} - V_{\mathrm{IAS}_{(1)}} \sin\psi_{(1)} \\ \vdots \\ V_{\mathrm{G}_{(M)}} \sin\phi_{(M)} - V_{\mathrm{IAS}_{(M)}} \sin\psi_{(M)} \end{bmatrix} = \begin{bmatrix} 1 & 0 & \cos\psi_{(1)} \\ \vdots & \vdots & \vdots \\ 1 & 0 & \cos\psi_{(M)} \\ 0 & 1 & \sin\psi_{(1)} \\ \vdots & \vdots & \vdots \\ 0 & 1 & \sin\psi_{(M)} \end{bmatrix} \begin{bmatrix} V_{\mathrm{W_N}} \\ V_{\mathrm{W_E}} \\ \Delta V_n \end{bmatrix} \tag{22}$$

to compute ΔV (and the wind, which is discarded) at the prescribed $\bar{V}_{\mathrm{IAS}_{(n)}}$ for each of the N test points. Then, by concatenating the results from the N test points into a single vector, one may simply use

$$\begin{bmatrix} \Delta V_{(1)} \\ \vdots \\ \Delta V_{(N)} \end{bmatrix} = \mathcal{M}_{\Delta}(\boldsymbol{\Phi}; \boldsymbol{w}) \tag{23}$$

to fit the SSPE curve. Again, this is an improvement over the ground speed course method because it generates higher fidelity estimates of each $\Delta V_{(n)}$, ultimately yielding a higher-fidelity estimate of the SSPE curve.

While methods of this type offer many benefits and makes better use of modern instrumentation, data acquisition equipment, and computational techniques than the ground speed course (and many other methods), there are still opportunities to streamline the method to further test efficiency.

2.3 A New Technique for Obtaining the Data

The focus of this paper is the turning acceleration method. The turning acceleration method uses a single test point conducted over the full operational range of V_{IAS} of the aircraft to fit a curve to ΔV over the entire range of V_{IAS}. During the test point, M measurements of V_{IAS}, V_{G}, ϕ and ψ are collected using high-speed data acquisition equipment (M is of order 10^3 or greater). Additionally, since the aircraft is maneuvered through a full turn, wind effects are observable and are incorporated into the model as an additional latent variable in the same manner as the cloverleaf and level turn methods. The model remains tolerant of wind noise. The key difference between this method and the previous method is that no assumptions is made that the test point is conducted at a fixed V_{IAS} (yielding a fixed ΔV), rather, V_{IAS} is assumed to vary during the test point. This yields many benefits, including increased tolerance of wind noise and execution error. Data reduction requires the full model described in (12) above since ΔV is no longer assumed to be constant throughout the test point. However, the entire SSPE curve may be calculated in one step, and from one test point.

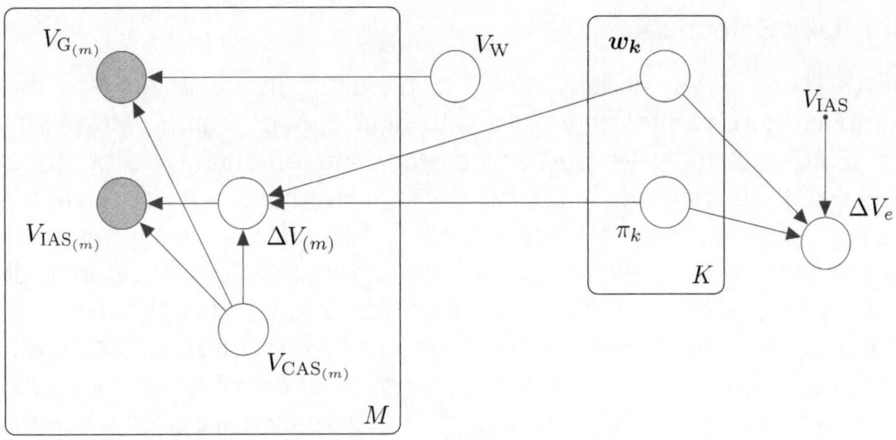

Figure 3: Graphical representation of the model structure and dependencies of the turning acceleration method.

Structurally, the model is similar to the cloverleaf and level turn methods, with the exception that V_{IAS} and ΔV are now free to vary throughout the test point. Also, several candidate models are fit to the test data and the final model is a mixing of the candidate models as described in the previous section.

3. FLIGHT TEST TECHNIQUE

Although the proposed algorithm and FTT were developed in support of a C-130J flight test program, the concepts, mathematics, and flight maneuvers necessary to perform the calibration algorithm apply to all aircraft. In general, the flight maneuver is as simple as flying the aircraft on a constant bank, coordinated 360 degree accelerating turn while maintaining altitude. It is important to note that sideslip directly introduces error into the algorithm since it corrupts the trigonometric relation between the heading (ψ), ground track (ϕ), and wind direction (φ_{W}) angles. The remaining details needed to successfully fly the technique are discussed below.

3.1 Altitude Selection

All airspeed calibration techniques rely on the adequate comparison of the TAS, GS, and wind vectors. As such, all airspeed calibration techniques, including this one, are ideally flown as close as possible to sea level (where Knots Indicated Airspeed (KIAS) and TAS are approximately equal) or at higher altitudes with additional computational and instrumentation requirements in order to calculate TAS from KIAS. Furthermore, the altitude tolerance required during the turn should be based on the desired confidence interval on the final correction curve. Using a general rule of thumb derived from cursory model sensitivity analysis, the maneuver tolerance should be $\pm$50 feet in order to assure a confidence interval no greater than $\pm$ 0.5 knots.

3.2 Angle of Bank Selection

Since Angle of Bank (AOB) directly drives turn radius, its selection may inevitably be driven by airspace constraints, especially with high speed aircraft. However, there are key considerations that must be addressed when implementing the proposed FTT for a particular aircraft. In general, most airspeed calibrations are conducted in order to generate configuration-dependent airspeed correction curves for aircraft flying at wings level. Since the error in airspeed measurement is generated, in large, by a disturbance of the static field around the aircraft, the factors affecting the disturbance characteristics should be carefully examined. Specifically, the static field disturbance has been shown to be a function of airspeed (and by corollary, a function of Angle of Attack (AOA) and load factor) [1]. Therefore, it is important to balance the airspace restrictions with the changes in AOA and load factor as a functions of airspeed, which are induced by changes in AOB. As such, it is recommended that the proposed FTT be flown at the lowest practical AOB in order to maintain the AOA-airspeed relation as close as possible to that of a wings-level aircraft. The mathematical algorithms described in this paper will yield a valid solution regardless of the angle of bank selected. However, the resulting correction curve may not necessarily match the curve of a wings-level aircraft if AOB selection significantly increased the AOA for a given airspeed.

3.3 Acceleration Profile

Similar to the method used in basic algebra to solve a system of linear equations, the statistical regression algorithm used in the proposed FTT (as well as the Level Turn and Cloverleaf) works by converging on a solution (a set of unknowns) given enough valid measurements (independent equations). The quality of the measurements provided to the algorithm directly affects the confidence interval of the final solution. By executing at least one full turn, the algorithm is given enough independent measurements to converge on the wind speed and direction unknowns. Similarly, by providing a wide range of airspeeds, the algorithm can then converge on the airspeed correction unknowns. As a result, it is recommended that the proposed FTT be flown using the widest possible range of airspeeds for a given gear and flap configuration profile. It is important to note that extrapolation beyond the data bounds of a particular model is seldom valid or accurate. Therefore, the airspeed range used during the execution of the proposed FTT should also take into consideration the calibration curve requirements driving the test (i.e. the range of airspeeds desired for the final calibration curve). Furthermore, in order to minimize the effects of wind variability on the measurement collection, the acceleration profile should remain as constant as possible. However, if airspace restrictions and aircraft-specific operating limitations yield an airspeed profile that is fully accelerated prior to completing a full turn, it is recommended that the remainder of the turn be completed at the last obtained airspeed. Finally, the current version of the algorithm is unable to model piecewise functions such as the ones created by transitions into transonic and supersonic regions. Therefore, a separate turning acceleration test point should be performed for each major airspeed region of the desired correction curve.

3.4 Instrumentation Data Rate

Although modern instrumentation systems are capable of recording at high data rates, it is important to carefully investigate the relationship between the recording system's sampling rate and the physical properties of the variables being recorded. That is, the rate of change of recorded variables should be analyzed and matched to the number of unique (independent) samples available at such rate. If any particular variable of interest changes at a slower rate than the number of samples recorded per second, the resulting repeated measurements allowed into the algorithm will falsely improve the Confidence interval (CI) of the results. In this particular case, the authors noticed the available data rate (100 Hz) far surpassed the Pitot-static system's ability to detect and provide independent samples. As such, prior to running the algorithm, the recorded data were filtered (yielding an effective sampling rate of 16 Hz) in order to remove repeated values that would otherwise artificially improve the confidence interval of the results. As later shown in Figure 10, the CI resulting from the filtered data was 0.88 kts while the unfiltered data yielded a CI of 0.35 kts.

4. RESULTS AND ANALYSIS

The new data reduction framework and flight test technique proposed by this paper were evaluated using numerical simulation and demonstrated using real-world flight test data. This section describes the findings from both.

4.1 Numerical Results

In order to determine the usefulness of the turning acceleration method as an FTT, numerical simulations of three aircraft were conducted using their flight manual SSPE curves as truth sources. An acceleration profile in CAS was prescribed for each aircraft, and the flight manual SSPE curve was then used to generate the IAS profile that would be seen in the aircraft and recorded by the data acquisition system. In order to demonstrate the algorithm's stability against variable or gusty wind, a mean wind component with uniformly distributed additive noise was added to the airspeed data. Finally, the variable wind was added to the CAS track in order to create the measured GPS track for the aircraft. The simulated data for V_{IAS}, V_{G}, ϕ, and ψ, was processed using the methods described in this paper in an attempt to recreate the truth data V_{CAS} and the SSPE curve (ΔV) obtained from the aircraft's flight manual.

Three aircraft with distinctly different SSPE curves were chosen in order to consider a range of possible outcomes. Simulated data were processed for each aircraft using a range of wind velocities and variability. A sample is provided for each aircraft as the first of two figures. Then, the velocity correction curve and associated confidence interval was produced for a single realization of the simulated data as shown in the left-hand plot of the second figure for each aircraft. Finally, a Monte-Carlo experiment of 1000 runs each for a set range of wind gust intensities was conducted to determine the influence of gusts on model confidence intervals.

4.1.1 Cessna 172

The Cessna 172 is a small, piston-driven, high-wing aircraft. A range of airspeeds from 40 to 120 Knots Calibrated Airspeed (KCAS) was selected to generate simulated IAS and GS data from a prescribed, smooth CAS acceleration profile using the flight manual velocity correction curve. A ground track and velocity polar are shown in Figure 4.

The left side of Figure 5 shows the estimated velocity correction curve with confidence intervals overlaid on the flight manual velocity correction curve. All candidate models were fit (quadratic, cubic, power, and exponential), and the power model was selected with an Akaike weight π of 1, indicating that the other models were considered non-informative by the data reduction algorithm and discarded for this particular realization of additive wind noise. It is clear that the turning acceleration method provides a good fit to the flight manual error correction curve and is capable of selecting the correct model type in this case.

A Monte-Carlo experiment was then conducted, generating 1000 data sets each for a range of gust intensities between 0 and 10 knots. For each data set, the velocity correction curve was recomputed. The average and variance of the 1000 confidence intervals was then determined for each gust intensity. The results are shown on the right side of Figure 5. This chart can be used to estimate an acceptable level of gust intensity for a specified tolerance for the velocity correction curve confidence intervals; for example, if one desires a confidence interval of 1 knot, test wind gust limits should be set to approximately 5 knots.

4.1.2 Lockheed C-130J

The Lockheed C-130J is a turboprop-driven, high-wing, intra-theater transport aircraft. A range of airspeeds from 140 knots through 260 knots was selected to generate simulated IAS and GS data from a similar prescribed, smooth CAS profile using the flight manual velocity correction curve. A ground track and velocity polar are shown in Figure 6.

The left side of Figure 7 shows the estimated velocity correction curve with confidence intervals overlaid on the flight manual velocity correction curve. All candidate models were fit (quadratic, cubic, power, and exponential), and a mixture model was selected, indicating that none of the models had an Akaike weight $pi > 0.7$ and that a mixture of several models provided a better fit than a single model for this particular realization of additive wind noise. The goodness of fit is similar to the C172 curve (note the y-scale of the graph).

The same Monte-Carlo experiment was then conducted, and the results are shown on the right side of Figure 7. This chart can be used to estimate an acceptable level of gust intensity for a specified tolerance for the velocity correction curve confidence intervals; for example, if one desires a confidence interval of 1 knot, test wind gust limits should be set to approximately 4 knots.

4.1.3 Boeing C-17A

The Boeing C-17A is a turbofan-driven, high-wing, inter-theater transport aircraft. A range of airspeeds from 100 knots through 300 knots was selected to generate simulated IAS and GS data from a similar prescribed, smooth CAS profile using the flight manual velocity correction curve. A ground track and velocity polar are shown in Figure 8.

The left side of Figure 9 shows the estimated velocity correction curve with confidence intervals overlaid on the flight manual velocity correction curve. All candidate models were fit (quadratic, cubic, power, and exponential), and a mixture model was selected, indicating that none of the models had an Akaike weight $\pi > 0.7$ and that a mixture of several models provided a better fit than a single model for this particular realization of additive wind noise. The goodness of fit is similar to the C172 and C-130J curves (note the y-scale of the graph).

The same Monte-Carlo experiment was also conducted, and the results are shown on the right side of Figure 9. This chart can be used to estimate an acceptable level of gust intensity for a specified tolerance for the velocity correction curve confidence intervals; for example, if one desires a confidence interval of 1 knot, test wind gust limits should be set to approximately 5 knots.

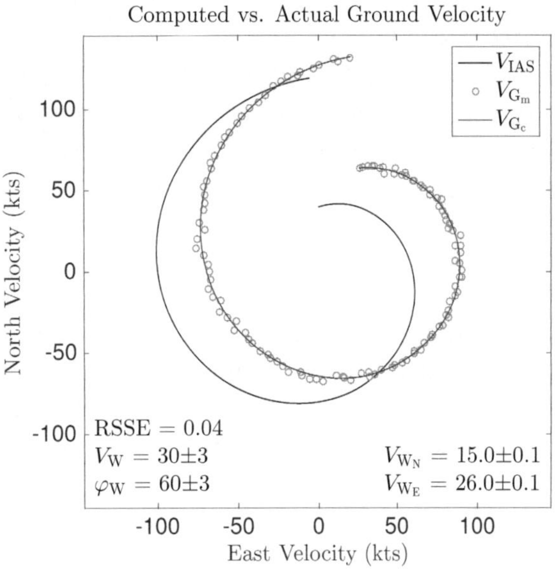

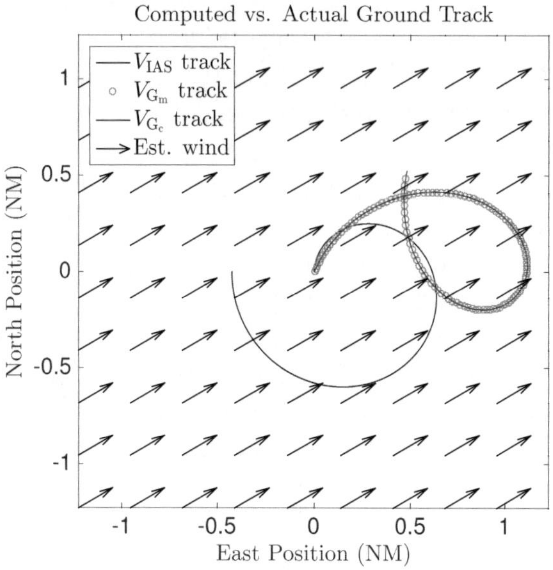

Figure 4: Cessna 172 Parameter Comparison. Left: Comparison of indicated airspeed (V_{IAS}), measured GPS ground speed (V_{G_m}), and computed ground speed (V_{G_c}). Right: Comparison of ground tracks derived from measured true airspeed (V_{IAS}), measured GPS ground speed (V_{G_m}), and computed ground speed (V_{G_c}) with computed wind vector.

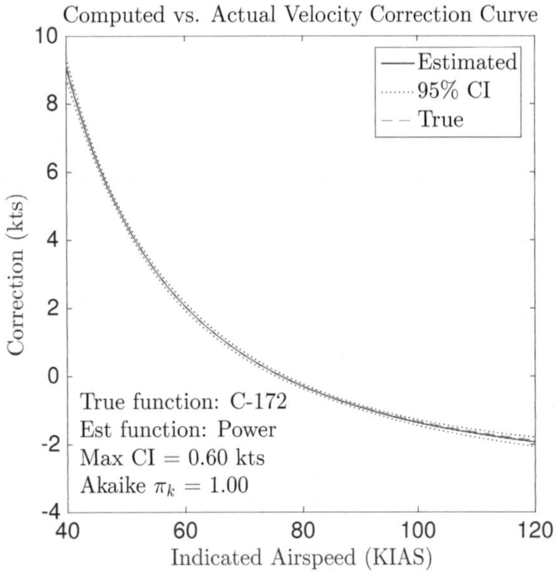

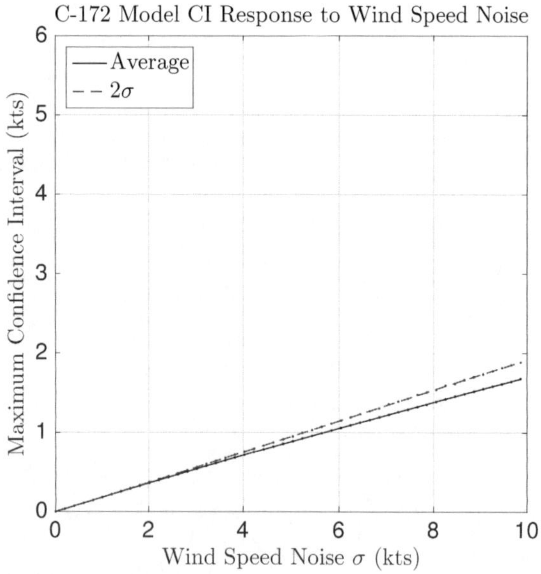

Figure 5: Cessna 172 Model Results. Left: Sample numerical results with 3 knots and 3 degrees of uniformly-distributed wind speed and wind direction noise. Model output with 95% confidence interval is shown to bound the True or known function. Right: Monte-Carlo results of 1000-run confidence interval response to increasing wind speed noise. Average maximum confidence interval and associated 2σ grow linearly with increasing wind noise.

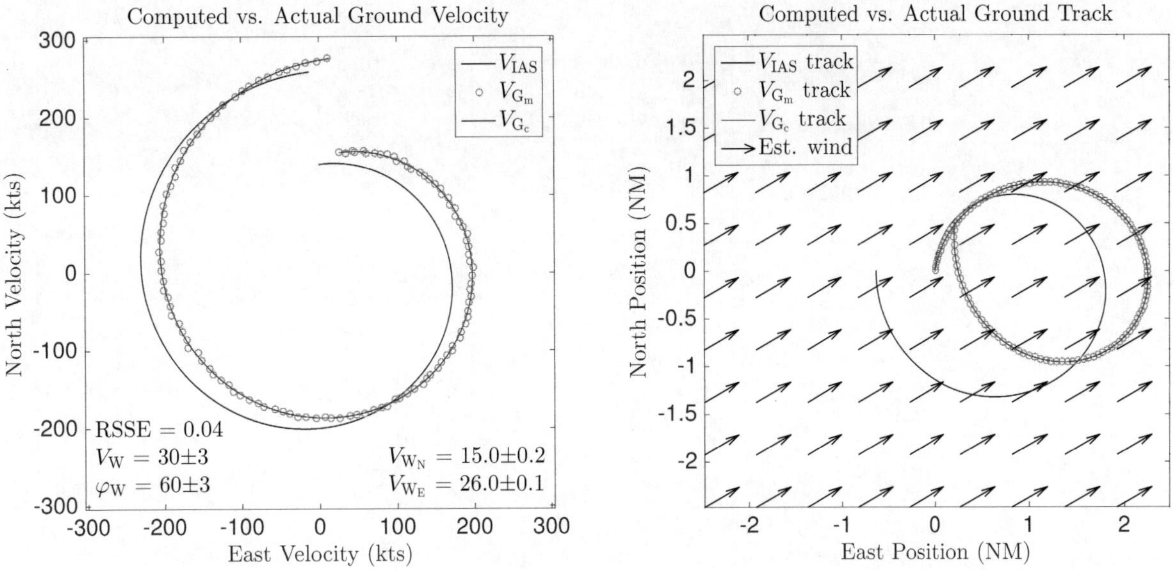

Figure 6: C-130J Parameter Comparison. Left: Comparison of indicated airspeed (V_{IAS}), measured GPS ground speed (V_{G_m}), and computed ground speed (V_{G_c}). Right: Comparison of ground tracks derived from measured true airspeed (V_{IAS}), measured GPS ground speed (V_{G_m}), and computed ground speed (V_{G_c}) with computed wind vector.

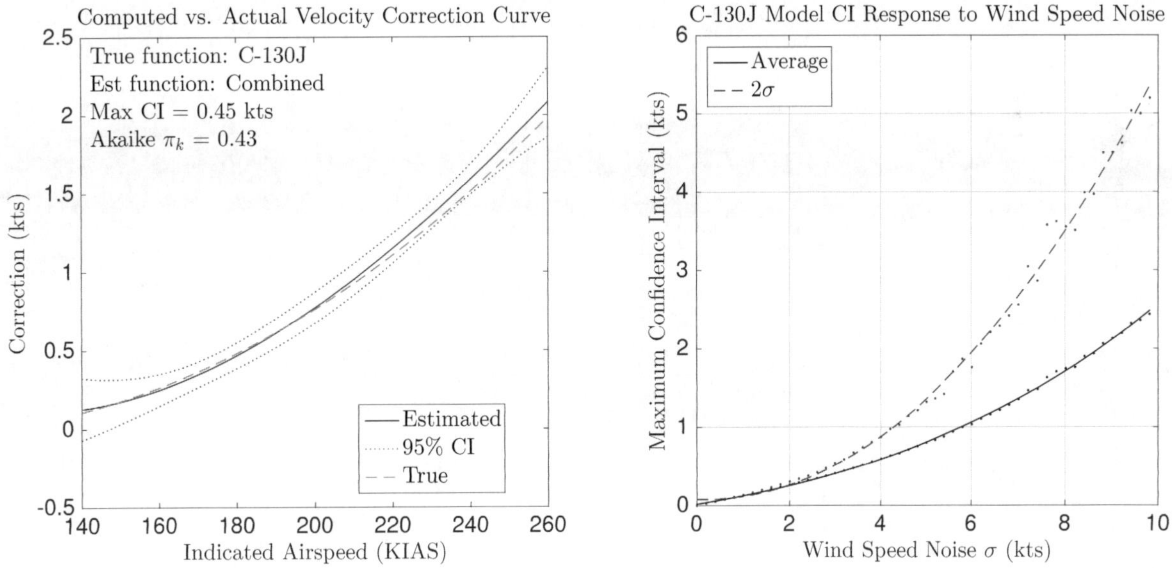

Figure 7: C-130J Model Results. Left: Sample numerical results with 3 knots and 3 degrees of uniformly-distributed wind speed and wind direction noise. Model output with 95% confidence interval is shown to bound the True or known function. Right: Monte-Carlo results of 1000-run confidence interval response to increasing wind speed noise. Average maximum confidence interval and associated 2σ grow linearly with increasing wind noise.

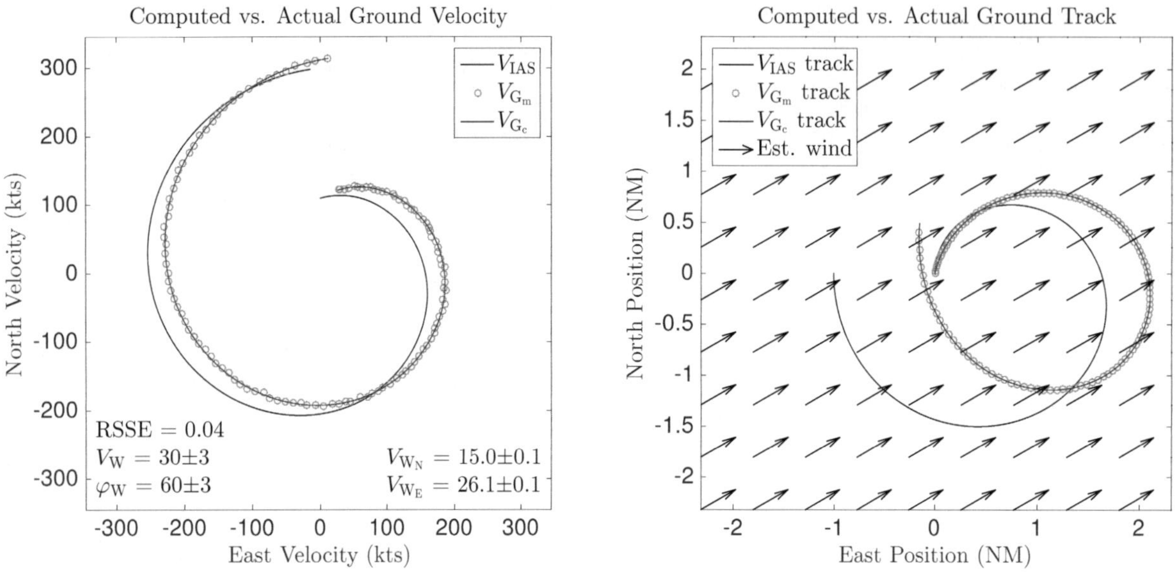

Figure 8: C-17A Parameter Comparison. Left: Comparison of indicated airspeed (V_{IAS}), measured GPS ground speed ($V_{\mathrm{G_m}}$), and computed ground speed ($V_{\mathrm{G_c}}$). Right: Comparison of ground tracks derived from measured true airspeed (V_{IAS}), measured GPS ground speed ($V_{\mathrm{G_m}}$), and computed ground speed ($V_{\mathrm{G_c}}$) with computed wind vector.

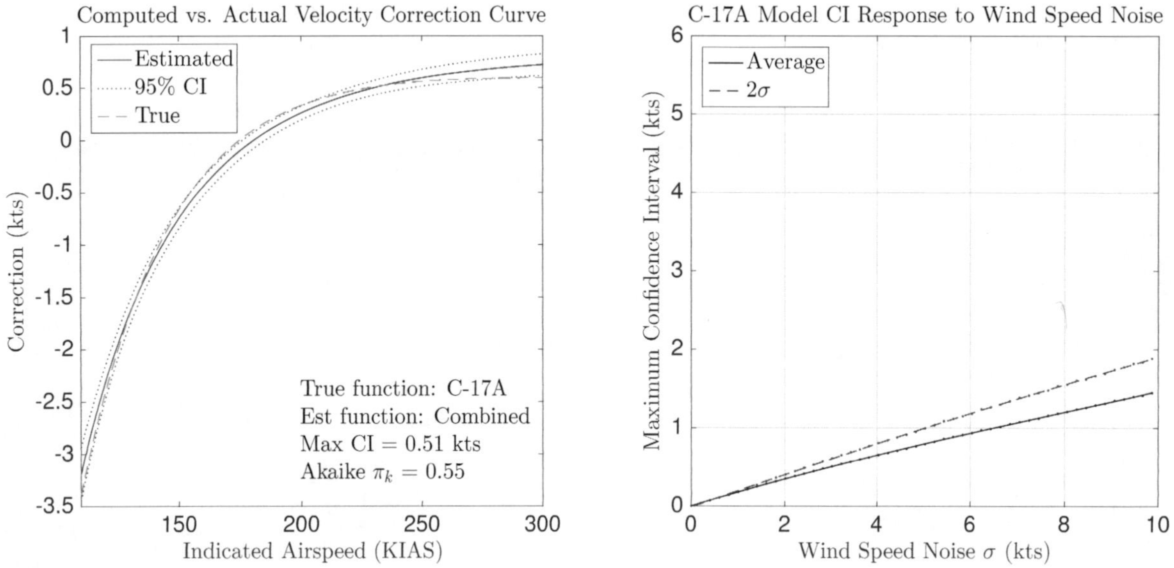

Figure 9: C-17A Model Results. Left: Sample numerical results with 3 knots and 3 degrees of uniformly-distributed wind speed and wind direction noise. Model output with 95% confidence interval is shown to bound the True or known function. Right: Monte-Carlo results of 1000-run confidence interval response to increasing wind speed noise. Average maximum confidence interval and associated 2σ grow linearly with increasing wind noise.

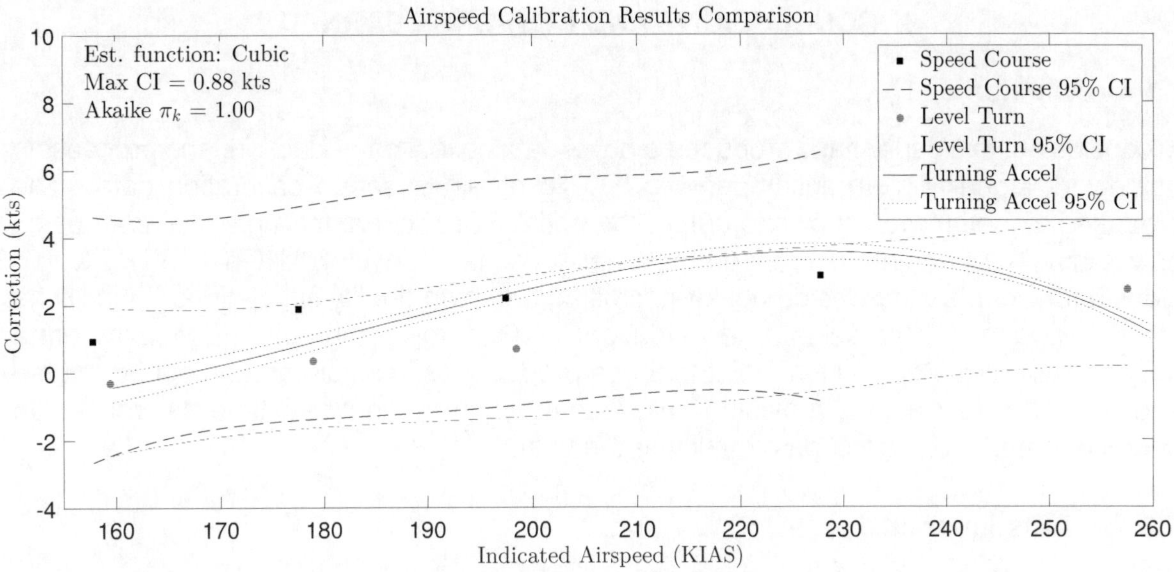

Figure 10: Comparison of airspeed calibration results from Turning Acceleration, Level Turn, and Ground Speed Course methods. The Turning Acceleration curve obtained using a single test point is shown to fall within the confidence interval of the other two methods, each obtained using 4 test points.

4.2 Real-World Flight Test Results

In order to validate the proposed algorithm, the proposed FTT was flown in along with the ground speed course and the level turn to calibrate a wingtip air data boom. As shown in Figure 10, the SSPE curve obtained using a single Turning Acceleration test point falls well within the confidence intervals of the curves obtained by the other two methods, each requiring 4 test points. Table 2 summarizes the flight time savings of the Turning Acceleration and Level Turn FTTs compared to the Ground Speed Course.

Table 2: Flight test time comparison.

FTT	Test Points	Flight Time	Savings
Ground Speed Course	4	36.5 min.	-
Level Turn	4	15.7 min.	57%
Turning Acceleration	1	4.6 min.	87%

As is evident from Table 2, the Turning Acceleration method offers significant time savings when compared to other existing methods. In fact, a sufficiently sophisticated data acquisition system could allow an air data system be calibrated in flight if it used a real-time implementation of the method in only a few minutes. This could allow a test team to recalibrate their Pitot-static system every flight, if so desired.

5. CONCLUSIONS AND LESSONS LEARNED

5.1 Conclusion

In conclusion, this paper has introduced a novel data collection, reduction, and processing framework along with an accompanying FTT to obtain airspeed calibration data. The proposed algorithm was shown to output the entire SSPE curve for a given aircraft using only a single test point. The algorithm's ability to handle widely differing SSPE curve types and variable wind was demonstrated using tree high-fidelity simulations. Finally, the algorithm was validated using real-world flight test data by comparing the results from a single turning acceleration test point to those produced by multiple ground speed course and level turn test points, providing an 87% reduction in flight test time as well as an unprecedented measure of data quality to the user.

5.2 Lessons Learned

The authors consider this proposed FTT and associated data management framework a lesson learned in and of itself. Nonetheless, three additional project-level lessons learned are identified and expounded on below.

5.2.1 Automation Limitations

During test procedure development for the real-world data collection flight on the C-130J, the authors and test team relied heavily on the use of C-130J automation (altitude hold, AOB hold, and auto-throttle) in order to minimize pilot workload and ensure an uniform acceleration profile. However, the team initially failed to realize the limitations of the C-130J autopilot in maintaining a coordinated turn (i.e. minimizing sideslip) while accelerating. Since the trigonometric relationships developed in this paper did not take into account sideslip angles, the data collection technique was repeated during a latter portion of the flight with additional emphasis on pilot-in-the-loop turn coordination.

5.2.2 Changing Paradigms

It is important to note the origin of the proposed FTT was formed out of framing the Pitot-static calibration problem using a different paradigm: "Can we model the underlying parametric curve governing SSPE instead of modeling a single airspeed at a time?" This single shift in the way the team observed the problem at hand led to a significant advance in its solution. In short, the authors recommend taking the time to freely brainstorm and think outside the box when facing not only new challenges but also challenges that have been seemingly solved before.

5.2.3 Investing in Knowledge

Finally, the authors would be remiss if the organizational leadership in charge of cost and schedule for the flight test program that motivated this research was not acknowledged for their support, trust, and investment in a possible better way of doing business. That is, the people ultimately responsible for executing the driving flight test program could have simply decided to perform Pitot-static calibrations using the well-known classical techniques. Instead, they decided to trust the research team and give them the necessary time and cost flexibility to not only come up with a better method but validate it using their real-world program.

REFERENCES

[1] R. E. Erb, Pitot-statics textbook, Tech. rep., USAF Test Pilot School (2010).

[2] G. V. Lewis, A flight test technique using gps for position error correction testing, in: COCKPIT, Society of Experimental Test Pilots Quaterly Publication, 1997.

[3] W. M. Olson, Pitot-static calibrations using a gps multi-track method, in: 28th SFTE Symposium, 1998.

[4] T. R. Jorris, M. M. Ramos, R. E. Erb, R. K. Woolf, Statistical pitot-static calibration technique using turns and self-survey method, in: 42nd International SFTE Symposium, Seattle, Washington, 2011.

[5] H. Akaike, A new look at the statistical model identification, in: IEEE Transactions on Automatic Control, Vol. AC-19, IEEE, 1974.

[6] K. P. Burnham, D. R. Anderson, Multimodel inference, in: Sociological Methods and Research, Vol. 33, Sage Publications, 2004.

Handling Qualities Stress Testing Revisited

K. Prosser (F)

Introduction

Flight testers have been in search of a flight test technique (FTT) that could be used to consistently uncover pilot in-the-loop oscillations (PIO) and other Handling Qualities (HQ) deficiencies for many years. The solution seems to have eluded the test community however. Despite numerous HQ Stress Testing techniques being introduced and published, none seems to have caught on to become universally applied. Many of these have been difficult to teach or difficult to understand or impossible to justify. Some have had no clear metric as to what is an acceptable and what is not an acceptable response. Despite the limited success some of the techniques demonstrated, a truly effective and acceptable technique must address all those limitations. To begin, none of these early HQ Stress Test techniques even attempted to define HQ Stress. An attempt to do that will be presented here and a potential HQ Stress Test technique will emerge that may prove more successful.

Stress Testing

Stress testing is a common technique used in flight testing to uncover deficiencies in various test items. During Auto-GCAS testing, stress testing was employed quite effectively. One of the favorite Auto-GCAS test locations was a small canyon in R2508 just northeast of R2505 called Rainbow Canyon (Figure 1). Rainbow Canyon was very desirable for stressing the system because it was a deep winding canyon that was extremely narrow. By flying through Rainbow Canyon, any latent GCAS deficiency would manifest itself as a nuisance flyup. If tested over flat terrain, a horizontal navigation error would not create a nuisance flyup. A scan pattern error would not cause a nuisance flyup, and a horizontal trajectory prediction error would not cause a nuisance flyup. In Rainbow Canyon, however, any deficiency in any direction would result in an unwanted nuisance flyup. Rainbow Canyon could, with one pass, uncover a great deal of information about the capability and limitations of the Auto-GCAS. What Rainbow Canyon could not be used for was an operational evaluation. Rainbow Canyon was not representative of most of the terrain in the world, and there is really no tactical utility in flying low level through it. There is no threat in the area, there are no targets of interest within miles, and there would be no way for a wingman to follow. Additionally, the canyon is too narrow to allow the pilot to perform any other required task while avoiding the canyon walls. A good Auto-GCAS may exhibit nuisance flyups in Rainbow Canyon and still be mission suitable. Rainbow Canyon was just a very efficient way to find out the capabilities and limitations of the Auto-GCAS. Handling Qualities can also benefit from a stress test but it is imperative that the distinction between a stress test and an operational evaluation be understood and appreciated. A HQ stress test, if it exists, could quickly and efficiently uncover HQ deficiencies and clear an envelope for other HQ testing but is not sufficient to determine mission suitability alone. For example, prior to conducting your first aerial refueling test, a HQ stress test could alert you to a possible PIO that, if encountered close to another aircraft, could be deadly. We would not use the stress test to evaluate the mission suitability of the aircraft, but only to 1) clear an envelope for other more dangerous HQ testing, and 2) efficiently determine if HQ limitations exist. The million dollar question is: does such a stress test for HQ exist?

Figure 1, Auto-GCAS F-16 flying through Rainbow Canyon[1]

History

The first significant milestone in the search for a HQ Stress test that this author is aware of was the Handling Qualities During Tracking (HQDT) FTT. It appears that the technique was first introduced sometime in the 1970s and called for the pilot to "aggressively and assiduously track a precision aimpoint." The technique was taught at the Air Force Test Pilot School for many years and ultimately was displaced by Boundary Avoidance Tracking (BAT) Workload Buildup (WLB). Table 1 shows some of the techniques that this author is aware of that have been introduced over time. So why have these techniques not garnered universal acceptance and use? There have been many critics of these techniques and they will point out that 1) they are difficult to teach, 2) there is often no build-up, 3) there is no metric for proper FTT execution, 4) there is no metric for goodness of the aircraft, and 5) there is no relationship between

[1] Photo courtesy Automatic Collision Avoidance Technology Program

the FTT and operational employment. The BAT WLB was a good attempt to address operational relevance. We could all appreciate that boundaries exist and we all understood our own behavior when in the presence of them. BAT WLB still did not meet with wide acceptance outside of academia and as we just noted, a stress test does not need to be operational relevant to be valuable. Confusing operational suitability evaluations with stress testing will lead to failure. In order to meet large scale acceptance, a HQ Stress Test FTT must have 4 characteristics:

1) It must be effective. In other words, it has to reveal latent HQ deficiencies in the airplanes that have them yet not indict an otherwise good aircraft.
2) It must have a measure of goodness. There must be some measure that distinguishes the good airplanes from the bad.
3) It must be repeatable and teachable. You should be able to get consistent results regardless of which pilot flies the technique. That means it should not require exceptional skill or elaborate training to perform.
4) It must be justifiable. We need more than a technique that works, but one that we all can understand why it works. If we do not understand why a technique works we cannot evaluate how well we performed it and we would be hopelessly confusing when we try to teach it. Note that this does not mean the maneuver flown has to be related to an operational task in some way. Pitch doublets, for example, have no operational counterpart but they are justifiable because we know what they do and how they work to answer flight test objectives.

TABLE 1 - Summary of Existing HQ Stress Tests

HQDT (~1970)	Aggressively and Assiduously Track a Precision Aimpoint
Aggressive Operational Tasks	Offset & Spot Landings, Fresnel Lens Approaches...
TPS HQDT (~1990s)	HQDT with Synchronous or Bang-Bang Endpoint
Clonk[2]	Full amplitude Synchronous Tracking on Pitch Rate
SIS PIO	Switching Induced Simulated PIO (another name for Synchronous or Bang-Bang control)
BAT WLB	BAT Workload Build-up through decreasing boundaries

What is HQ Stress?

[2] This has been referred to as "Klonk" and "Klunk" in some literature

It is this author's belief that many of the legacy Stress Test techniques failed to achieve large scale acceptance because they failed to meet the fourth criteria listed above. One could further argue that the reason they failed to meet that criteria is that none of the listed techniques defined HQ Stress. **How can we conduct a HQ Stress test if we do not know what HQ Stress is?** More importantly, how can we do a build-up in stress if we do not know what stress is or how to measure it? No doubt one or more readers are on the edge of their seat ready to proclaim the answer to the question of what HQ stress is. They are likely to say that it is "pilot gain." Assuming that is correct, what is pilot gain? We see pilot gain used everywhere in the context of HQ testing. The Committee on the Effects of Aircraft-Pilot Coupling on Flight Safety used pilot gain in its report "Aviation Safety and Pilot Control." The Mil-STD-1797 talks about it. The FAA Advisory Circular 25-7C definitively declares that increased pilot gain is a prerequisite for exposing PIO, but what is it? A clear definition of pilot gain seems to elude us. Ina Niewind wrote an interesting paper entitled "What the hell is Pilot Gain." In that paper, Ina says: ***"It is curious that even though pilot gain is one of the most important aspects in handling qualities testing and even though it is very well understood on an intuitive level, there is no generally accepted verbal or mathematical definition."*** Ina Niewind was in search of a metric to measure pilot gain when she made the above observation about the definition of pilot gain. Bandwidth, frequency content, pilot inceptor workload, duty cycle and many other metrics have been used to quantify pilot gain but none have emerged as the clear answer. The very fact that there are so many different measures of the same thing should raise concern about the applicability of pilot gain in general. Ina's paper prompted an informal investigation and survey of pilot gain. Students and peers were asked over the years to define pilot gain and, to date, there have not been consistent answer from any two pilots. We all seem to have a sense of what it is but we cannot define it. Respondents defined high gain with descriptors like aggressive, large, high frequency, high mental workload, high intensity, and high bandwidth among others. Ultimately, whatever gain is, it must lie somewhere in the universe of possible piloting techniques so it is worth briefly discussing that now. Any possible piloting technique will consist of three characteristics: inceptor size, inceptor rate, and some amount of compensation. The inceptor size and rate describes the physical work being performed but it is incomplete. The timing of pilot inputs and the shaping of inputs is as equally important to performing a closed loop task as the physical characteristics. The compensation element includes all of those things a pilot does to improve performance. It includes lead and lag compensation and it includes pre-emptive and precognitive inputs. Physical characteristics are easy to measure and quantify but the level of compensation is much harder if not impossible to do so. No attempt is made to quantify it at this point only to recognize that it is an element of all piloting techniques. Figure 2 illustrates the universe of piloting techniques. HQ Stress lies somewhere in that universe as does pilot gain. Notice that "boundaries" and "triggers' are nowhere to be found in this universe. Boundaries and triggers are stress to the pilot but not the aircraft. The aircraft is only aware of what the pilot inputs are but not the reason for those inputs. Boundaries and triggers will change how a pilot flies and may do so abruptly, but whatever that new pilot behavior is, it will lie somewhere in the universe of piloting inputs just described. When the responses to the informal gain study are distilled down to the three axis of the piloting universe, high gain falls in as large, fast, and high compensation. Not all the pilots would use all three of those elements but if they did use one, that is how it would fit in the piloting universe. A build-up in gain is illustrated in Figure 2 and notice that high gain does not reside in the high work corner of the universe. Since the piloting task is closed loop, it is impossible to force high gain into the high workload corner. As soon as a pilot begins to compensate, the pilot is compensating for a purpose (to improve track performance or boundary avoidance). Once the pilot begins to compensate, their inputs are constrained and arbitrarily large or fast inputs cannot be made. This behavior was proven by having pilots track targets with low gain and with high gain and recording the physical characteristics of their inceptor inputs. Figure 3 illustrates one of these tests. Notice that the high gain tracking involves a slight increase in stick deflection and rate over the low gain task but not nearly to the aircraft limits. Similar results were observed when avoiding boundaries as when tracking targets. The alternative is also true. Once a pilot begins to compensate, they are no longer free to put an arbitrarily small or slow input. As a result the build-up in gain is never as clean as shown in Figure 2. Instead it is very noisy and can have wild excursions at the mercy of the task being performed.

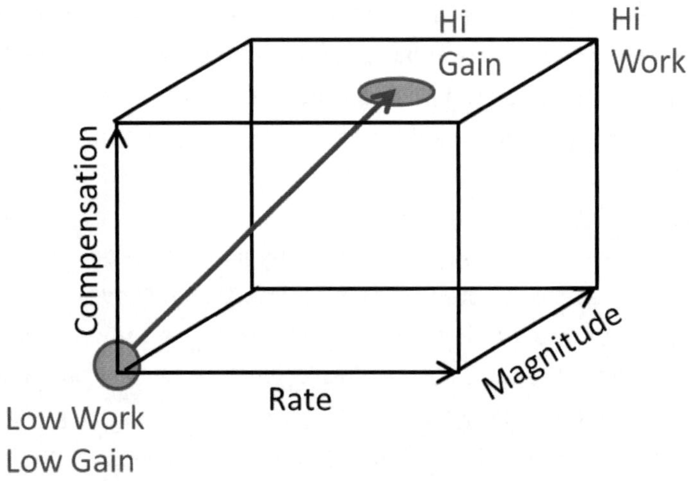

Figure 2, Piloting Universe with Gain Build up

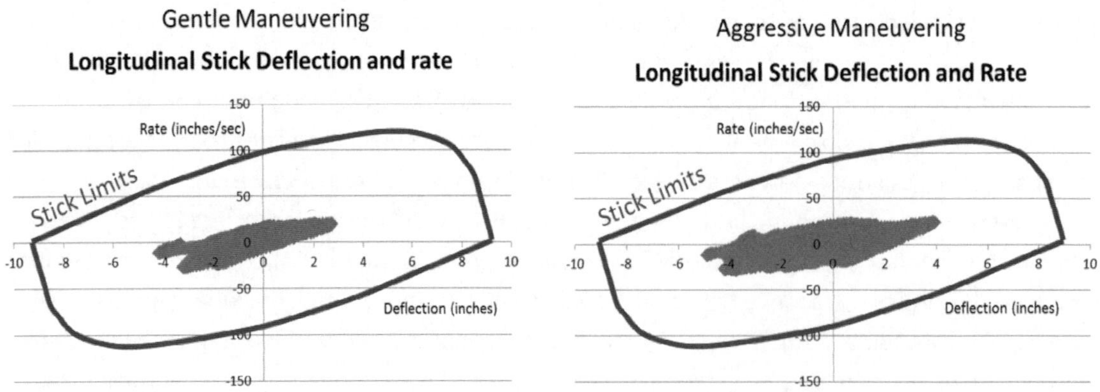

Figure 3, Pilot inputs during Low Gain and High Gain Tracking

Personal Experience

In 1999 through 2003, I taught at the USAF TPS as an active duty pilot and was the Flying Qualities Branch Chief at that time. The commandant of the school tasked me with conducting an evaluation of the HQDT technique. His concern was why we were teaching a technique that was not being received widely outside of academia. He wanted to determine once and for all, does it work and should we keep teaching it? As branch chief, the job fell on me and, with the help of many others including Andrew "Rocket" Thurling, we prepared and executed a test. The results were reported to the Society of Experimental Test Pilots in 2000 at the annual symposium. In a nutshell, the technique worked. Unfortunately it was difficult to teach and there was not a clear metric as to what goodness was and there was no accompanying justification for it. Without an alternative, the recommendation was to continue to teach it until something better came along. With a little bit of experience, I can look back now and see why HQDT encountered problems. Figure 4 illustrates the techniques as it was taught at the school when I was on staff in 1999.

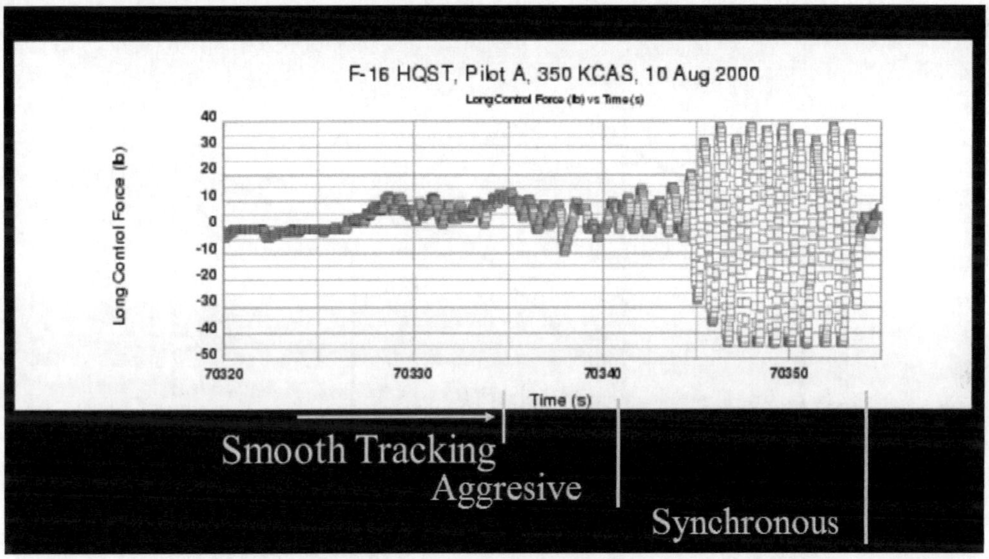

Figure 4, HQDT as taught at the USAF Test Pilot School in 1999

The technique began with a build-up in tracking aggressiveness. In other words, a build-down in allowable track error to the point that eventually the pilot would be tracking aggressively and assiduously while correcting for even the smallest error. In this phase, pilot "gain" was increasing. Inputs were more abrupt and compensation increased dramatically. Control displacement did not usually increase dramatically however for the reason we discussed earlier. **Compensation has to have a purpose**! As a result, when compensation increased, pilot inputs would get sharp (faster) but the size of the inputs would not grow significantly. Once the pilot achieved the most aggressive tracking and employed the absolute maximum compensation to minimize error to the utmost, the technique culminated in a switch to a synchronous tracking technique wherein any error was corrected for with a fixed step input. When the target was overshot, a quick reversal would be made making it a closed loop square wave or what was called "bang-bang" control. The size of the pilot's input would be increased until full deflection was achieved if the aircraft limits allowed. I do not believe that the synchronous control endpoint was part of the original HQDT technique but instead evolved over time. Nevertheless, by 1999, this was the technique being taught at the USAF Test Pilot School and the technique that was evaluated in the 2000 flight test[3]. The first half of the technique was a build-up in gain through a build-up in compensation. In other words, the pilot worked harder and harder to achieve more and more accurate track performance. Performance in fact that went way beyond any mission requirement, but that was alright because this was a stress test and not an operational suitability evaluation. The second half of the technique, the synchronous portion, actually had zero compensation. McRuer[4] described synchronous tracking in his HQ research as a zero compensation zero time delay tracking methodology. The pilot does not add lead or lag compensation and no stick shaping takes place. The TPS HQDT technique is mapped to the piloting universe in Figure 5. So what was the build-up in? Was it stress? Is more compensation the same as more stress or is less? Most pilots would say more compensation is more "pilot gain" but is it more stress? I believe this technique failed to achieve wide acceptance because it did not answer the fundamental question of what stress was and what is used to measure it.

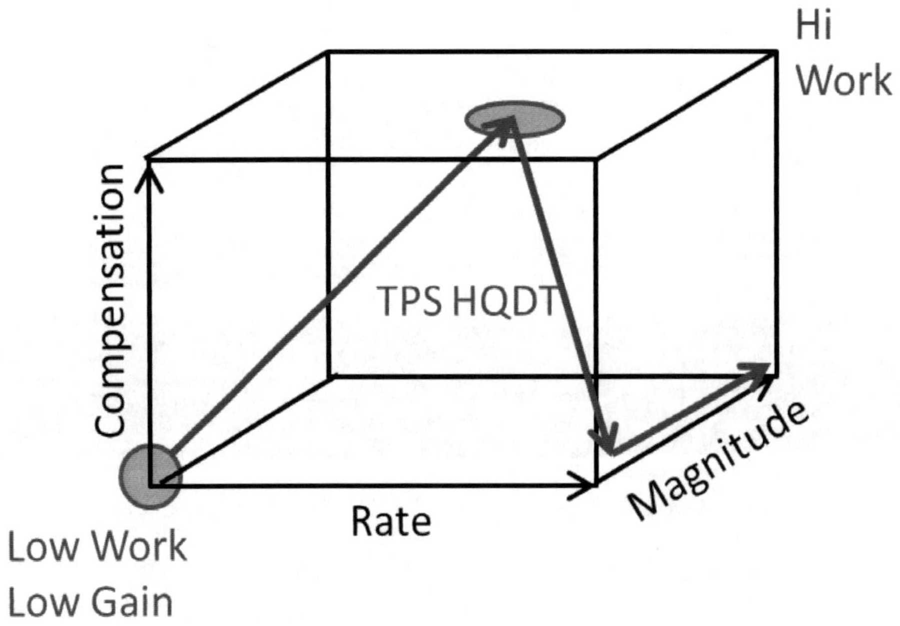

Figure 5, TPS HQDT Technique within the Piloting Universe

[3] In 2006 the Bang-Bang control strategy would come to be called "SIS PIO" for Switching Induced Simulated PIO
[4] McRuer, "Pilot-Induced Oscillations and Human Dynamics Behavior" NASA Contractor Report 4683

HQ STRESS Definition

In order to define HQ Stress, we need to take a step back and define why we want to know what it is. Simply put, we want to know what HQ Stress is because we believe that we can uncover latent HQ deficiencies when we stress an aircraft that we could not uncover without stress. Let us then use that as the definition of stress. **HQ Stress is, by definition, the piloting technique where latent HQ Deficiencies will manifest that do not manifest with other piloting behaviors**. "Latent HQ deficiencies" can be PIO but are not restricted to only PIO. With this definition we have a path to understanding HQ Stress. If we can understand where and why latent HQ deficiencies manifest we can establish what HQ Stress is. In Rogers Smith's Handling Qualities Review[5] presentation, Rogers' identified 5 ingredients of a PIO (arguably this applies to all bad HQ characteristics). The 5 ingredients are listed below.

1) Pilot Plus Aircraft
2) High Gain & Closed Loop
3) Boundary Conditions
4) Latent HQ Deficiencies (to include time delay, rate limiting, non linearities, high control sensitivities and poor control dynamics)
5) Trigger.

These elements in various forms are repeated throughout the HQ literature. These ingredients form the outline of a HQ Stress test. Since it is the latent HQ Deficiencies we intend to find, we should be able to add the other ingredients and the deficiencies should manifest. The other ingredients are: Pilot Plus Aircraft, High Gain Closed Loop, Boundaries, and Triggers. Performing these four ingredients is the Stress Test. I think most pilots would agree with this synopsis of the current HQ literature but it still does not answer the question of what stress is because we are right back to "pilot gain" and we do not know what that is. Notice that Smith went a step further than his peers here however. He recognized that high gain tracking was not sufficient alone. Boundaries and triggers were necessary to reveal the latent defects. Boundaries and Triggers are not pilot inputs however, so the boundaries and triggers must act in some way to modify the high gain tracking. This modified high gain tracking is our Stress Test but unfortunately it remains undefined. There have been some attempts to correlate an engineering metric to "pilot gain." The Mil-Std-1797, for example, defined high gain as "wide bandwidth". Bandwidth is something we can measure but is it stress? I have also seen duty cycle and its cousin Pilot Inceptor Workload (PIW) as a definition of pilot gain. Even mathematical gain has been used to define gain, but again, are they proper surrogates for HQ Stress? Let's go a little farther back and look at the raw data. Figure 6 below illustrate the pilot inputs and pitch rate response during the famous YF-22 PIO. Going back to our definition of HQ Stress as the piloting technique when bad things happen, we can probably agree that a bad thing was happening. Ultimately the aircraft was damaged beyond repair and never flew again. The pilot's inputs are in the top strip and the aircraft pitch rate response in the bottom strip. To the left, the aircraft is flying nicely and to the right a PIO (Latent HQ Deficiency) is in progress. Let's look at how the pilot inputs change from the time when things are ok (left side) until he is in a full-fledged PIO (right side.) Notice that as the aircraft pitch rate begins to diverge, the pilot's stick inputs grow. We could say his gain (mathematical

[5] Handling Qualities Review...a Practical Perspective, presented Manching Germany 11 Nov 2008 by Rogers Smith

gain) is increasing but, technically, that would be inaccurate. The pilot's inputs are indeed growing, but they are doing so in response to an ever growing pitch error. Arguably his gain is pretty constant. We could look at duty cycle, but notice he is in the loop continuously prior to the PIO as evidenced by the continuously changing inputs. We can see that once the PIO develops, there are many times where the input is fixed either at the stop or short of it. The pilot inputs begin to look like the synchronous bang-bang technique described previously. The pilot's duty cycle actually drops significantly while in the PIO. For that reason, his bandwidth also drops as almost all of his energy is DC while holding the stick at a stop. There is some high frequency component but it is of short duration so very little total power exists at high frequency. What does change? The rate at which the pilot moves the stick and the magnitude of the deflections increase dramatically when in the fully developed PIO. The YF-22 PIO is actually a very complex PIO because a mode change occurs and the aircraft in the PIO is not the same as the pre-PIO aircraft so a full review of other PIOs is in order. Figures 7 through 8 illustrate a couple of the many other PIOs that were reviewed and all are characterized by rapid and often full deflection stick inputs.

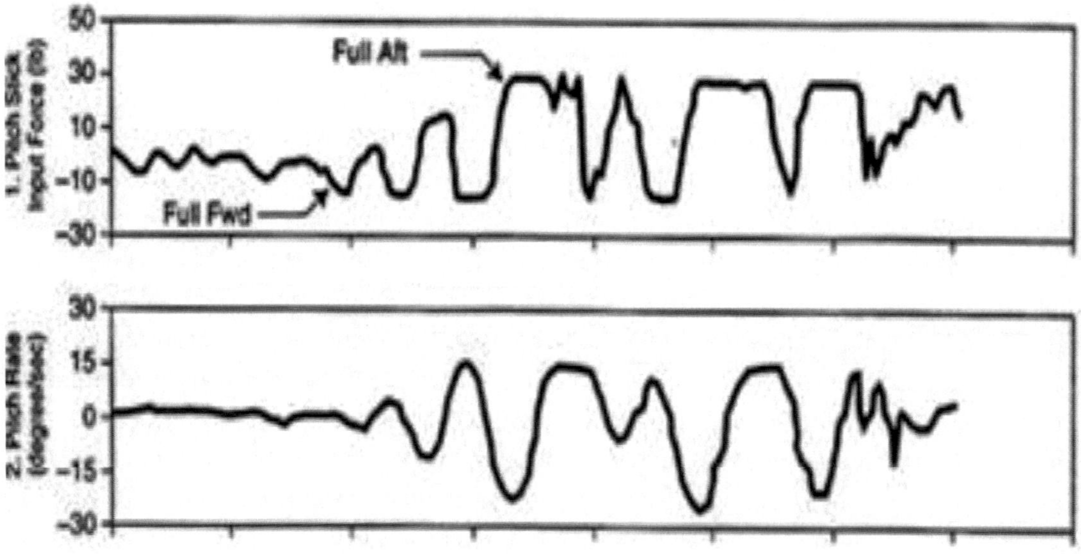

Figure 6, YF-22 PIO[6]

---

[6] PIO strip charts taken from Data: "Aviation Safety and Pilot Control" -Committee on the Effects of Aircraft-Pilot Coupling on Flight Safety

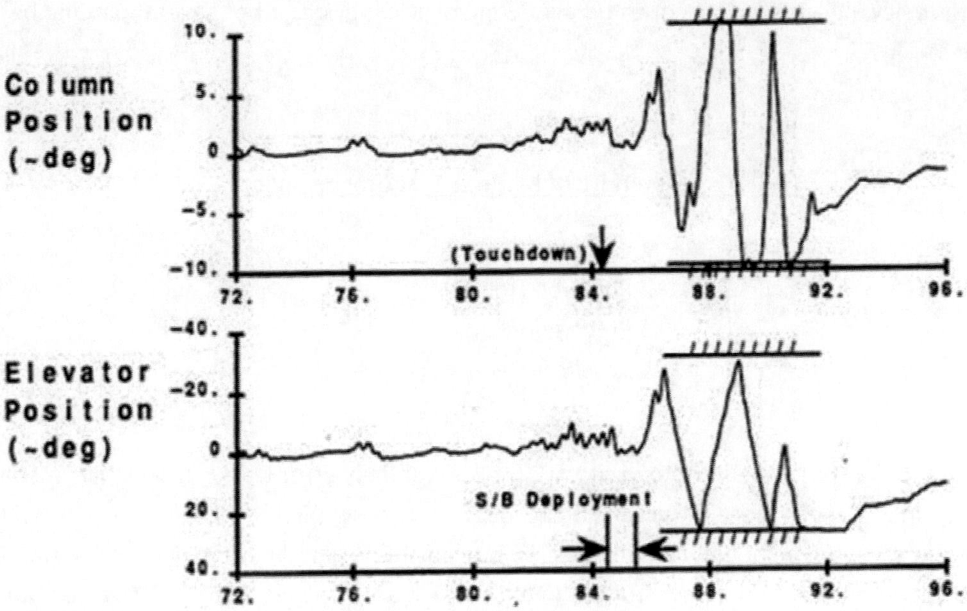

Figure 7, Large Aircraft PIO

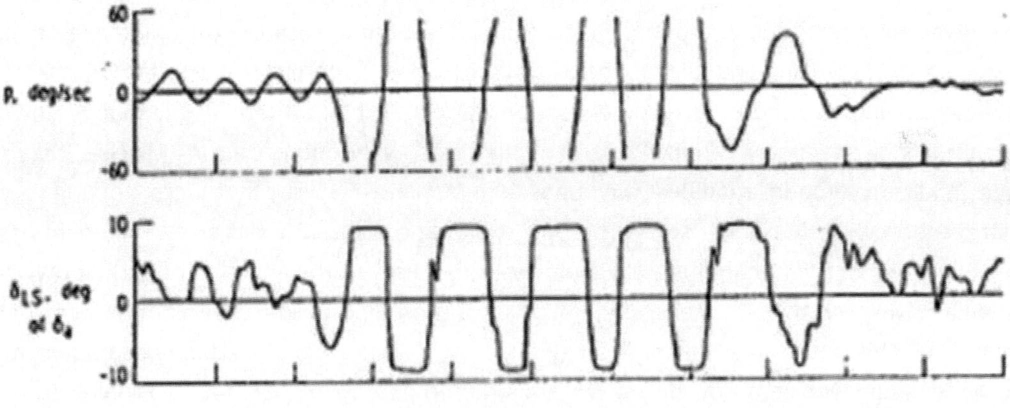

Figure 8, M2F2 PIO

Historical PIO evidence suggests to us that the traditional measures of pilot gain like mathematical gain, bandwidth, and duty cycle are not indicative of HQ Stress. What we do see is that control inputs get rapid and they get big in fully developed PIOs. If bad things happen when closed-loop control inputs are rapid and big, we have found our definition of HQ Stress. Furthermore, the synchronous bang-bang control strategy discussed previously is often observed when bad HQ are experienced. The synchronous control strategy was noted by McRuer in “Pilot-Induced Oscillations and Human Dynamics Behavior.” It may be worth reviewing McRuer's theories on pilot control. McRuer describes several piloting behaviors listed in table 2.

TABLE 2 – Summary of McRuer Tracking behaviors

1 Compensatory
2 Pursuit
3 Precognitive
4 Synchronous

The first behavior is called compensatory. This is essentially error tracking. The pilot responds to the error seen in whatever loop is tracked. If the pilot is tracking pitch attitude, the pilot will apply corrections to zero out the error between the desired pitch attitude and the actual pitch attitude. You can think of this as the normal human control technique. It is true for flying airplanes but also for riding bicycles and coloring inside the lines while in kindergarten. Pursuit occurs in conjunction with compensatory but the pilot has "preview". In other words the pilot has some means to predict what the error will be in the future. Precognitive behavior is when the pilot, with enough experience, can predict and apply the precisely required control input to zero out the error in a preemptive and open loop manner. Synchronous behavior is often seen in PIOs and it seems that the pilot tracks with, what appears to be, zero time delay and zero compensation. Presumably this occurs because the pilot sees a pattern (consciously or unconsciously) and mirrors the aircraft response. Regardless of why the pilot tracks that way, the response is synchronous, zero compensation tracking. What is interesting is that when you talk to a pilot that has just been in a serious PIO and applied this synchronous tracking process, they do not believe they had zero compensation. Just the opposite in fact, most will say they were working very hard. But the strip charts will tell another story. The only explanation is that under stress, the pilot in a fully developed PIO is working very hard mentally and physically to overcome and eliminate their own normal response delay. They see this as extensive lead compensation employed to overcome an out of phase lagging response. This results in a control strategy that is actually devoid of lead or lag compensation (or any stick shaping) and with zero time delay. While a stressed pilot works very hard to achieve this, it can be replicated in flight test with relative ease. To mimic this behavior, the pilot puts in a fixed size input and holds that input until the error goes to zero (pipper crosses the target). Once the target is reached, the pilot quickly reverses the input with an equal size input in the opposite direction. Since the reversal occurs at zero error, an overshoot will occur and is not indicative of a deficiency. The process repeats with the pilot switching signs on the input signal at the zero crossings. The switching process is why the technique is sometimes called "Switching Induced Simulated PIO" (SIS PIO). With a little practice, it is actually quite easy to accomplish and, In fact, a relatively simple closed loop Programmed Test Input (PTI) has successfully been employed by the USAF Test Pilot School to automate the process in a closed loop manner.

If McRuer is correct, we can expect to see pilots employing any one of the above techniques whether consciously or not at various times. What is interesting is that the first three techniques describe tracking behavior with various levels of compensation employed to improve track performance, while the fourth demonstrates a strategy that is nearly devoid of what we would traditionally call compensation and is typical when bad things are happening. While the pilot is working hard, he is mimicking the error in a bang-bang like manner with little control shaping and no lead or lag compensation. Perhaps we should refine our definition of HQ Stress to be large, rapid control inputs with no compensation. Notice this is at odds with what most pilots call "gain" which is rapid inputs with maximum compensation.

Referring to Mil-HDBK-1797 we can glean a little more insight. In the Mil Handbook, the causes of PIO are listed. Among them is the statement that PIOs result from "excessive demands on pilot adaptation." With "required gain, lead, lag, lie outside the range of normal capabilities." And "rate of adaptation is too slow to preclude oscillations."[7] In other words, the PIO manifests when the pilot is unable to compensate enough to keep it hidden. Many readers might recall their first formation in the T-38 while in pilot training. The T-38 is known to be PIO prone and pilots often PIO on their first attempt at close formation flying. However, with experience they learn to adapt and to compensate and suppress the PIO tendencies and are eventually able to successfully complete the task. Here it seems that the latent HQ deficiencies (the PIO) is unleashed when the pilot cannot compensate enough to keep it hidden. If this is true globally, a very important concept emerges. If compensation can mask latent deficiencies, the lack of compensation can reveal them. If reducing compensation reveals HQ deficiencies, shouldn't our HQ Stress testing build-up be a **build-down in compensation**? Again we find our endpoint on the HQ Stress scale to be large, rapid inputs made with zero compensation. This HQ Stress end point is at odds with the traditional belief that we should build-up in pilot-gain through increasingly more aggressive compensation.

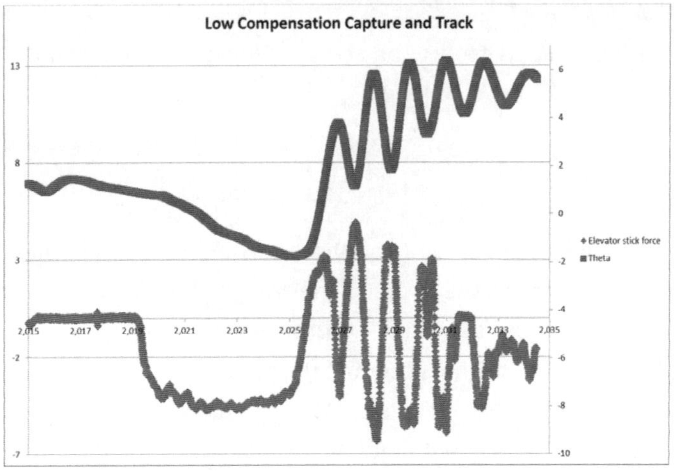

Figure 9, Pitch capture and track with low compensation

[7] Pg 273 MIL-HDBK-1797

Figure 9 and 10 show the pitch response and elevator stick inputs during a student tracking attempt in the Learjet Variable Stability In-flight Simulator during a USAF TPS demonstration. The configuration used was a low pitch damping configuration (damping ratio ~ 0.2) which is susceptible to PIO. The first trace is one student's first attempt at the task. With very little experience, the pilot applied limited compensation and a small amplitude PIO resulted that lasted several cycles. Pilot inputs at the time were large and rapid. The second plot shows the same student that applied learned compensation to improve performance. Notice that an incipient PIO begins at the capture but, through compensation, is quickly suppressed. The pilot would be applying "hi gain" or "hi compensation" piloting to improve performance and he succeeded in doing so. If this were a test to find deficiencies (the latent PIO tendency in this case), the "high gain" pilot would suppress it while the low compensating pilot would reveal it.

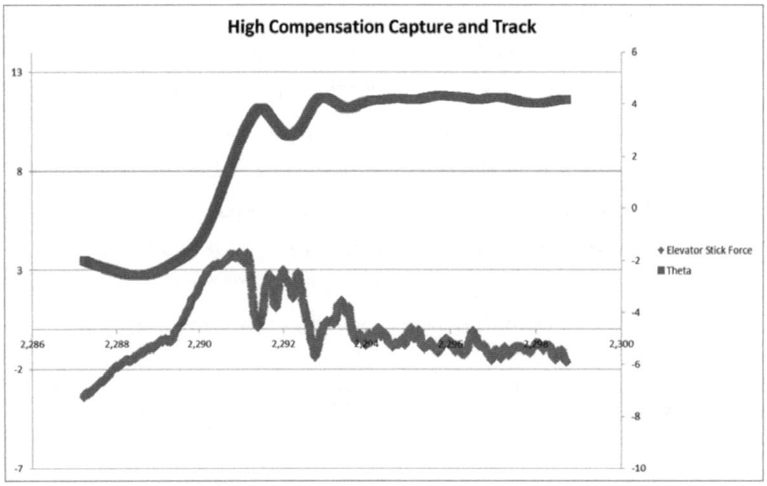

Figure 10, Pitch capture and track with high compensation

Boundaries and Triggers

In the preceding discussion, a very critical phase of the PIO was overlooked. We discussed the piloting technique that existed prior to the PIO and the piloting technique that was evident during the fully developed PIO but we did not discuss the very complex incipient phase of the PIO and the piloting technique that triggered the PIO. One might argue that the piloting technique that triggers the PIO is more "stressful" than the piloting technique that sustains it. There are a couple reasons why this is not true. The first is a logical reason. If the piloting technique that occurs after the PIO is triggered is not as bad as the piloting technique that triggered the PIO, why does the PIO grow and sustain? If the post trigger piloting technique was not as stressful, then the aircraft closed loop response should decay back to the pre-PIO response which it does not. The second reason is that we have a lot of experience with boundaries and triggers and we know what the piloting behavior is that triggers the PIO. In the extreme cases, when stressed by a trigger or boundary, a pilot reacts in a primal fashion. McRuer calls this a "transition from full-attention pursuit or compensatory operations in high-gain, high urgency tasks to a synchronous mode of behavior." Pilot's know it as a primal and instinctive knee-jerk like reaction to a threat. Control inputs become abrupt and without compensation. The point is that the piloting behavior that triggers the PIO is the SAME piloting behavior that sustains the PIO. In some PIO's, the pilot tracks a different error signal from the pre-trigger strategy. For example, prior to the PIO, the pilot may be tracking pitch attitude initially, but after the trigger, the pilot may sustain the PIO while tracking normal acceleration. No attempt is made here to understand or mimic those complex dynamics nor do they necessarily need to be. We simply recognize that a HQ Stress test may need to be conducted using several different error signals. A pitch attitude HQ Stress test followed by a pitch rate HQ Stress test for example.

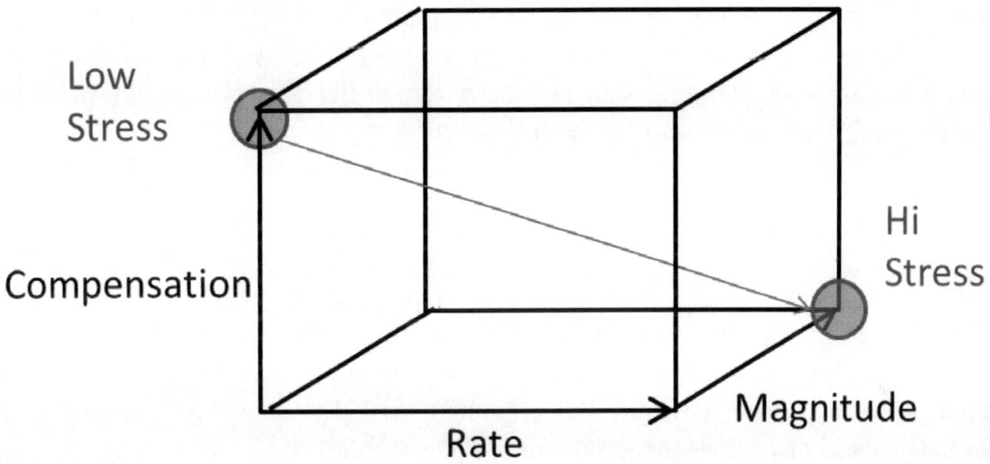

Figure 11, Piloting Universe with HQ Stress Build-up

Let's pretend for a moment that the above definition of stress is true. It would explain a lot. Notice that the high stress corner is precisely where the USAF TPS HQDT technique brought us. It is also where the Klonk technique resides. Those techniques seem to have found the high stress corner and their only failure lied in the lack of a justification, proper build up, and a metric in which to evaluate the goodness (or badness) of the aircraft. It may also help us understand pilot gain a little better. Notice that a build-up in gain, as shown in Figure 2, increases stress in an oblique fashion. This may explain why an increase in gain can often succeed in uncovering a latent defect and why it fails to do so at times. It may also explain why there are so many different definitions and techniques to accomplishing a gain buildup. Since it is not on a direct path of HQ Stress build-up, strict adherence to one type of gain build up has little advantage over another.

Non PIO HQ Deficiencies

The discussion so far has focused on PIO and not HQ deficiencies in general. The fact is that most of the published data on severe HQ deficiencies involve PIO and most would agree that a divergent PIO is certainly one of the most severe HQ deficiencies possible. This is not to mean to mean that this discussion and the technique to be presented are limited only to PIO detection. Poor HQ could be severe such as PIO but poor HQ can also simply be the lack of precision necessary to perform the required task. Precision and workload are critical elements of a HQ evaluation but they are also very mission dependent. For example, an aircraft that is not precise enough to perform rocket attacks may be perfectly suitable for airdrop missions. The purpose of HQ stress testing is not to determine mission suitability and an effective HQ stress test is neither designed to nor likely to provide that information. The purpose of the stress test is to clear an envelope and provide confidence to the evaluation pilot that the planned operational evaluation will be safe to perform. HQ deficiencies that a stress test must uncover include PIO, but also include cross axis coupling, stability and control issues, over-g potential, and over-control potential to name a few. While not proven here, the worst case piloting technique for PIO probably corresponds to the worst case piloting technique for each of the other potential problems just described. At this point we simply recognize that the piloting behavior in a PIO appears to be the worst case piloting behavior from a HQ Stress standpoint and leave the proof for later research.

The Technique

This brings us to the proposed HQ Stress test FTT. We want to build-up in stress to the point where the pilot's inputs are rapid and full deflection with no compensation. Low stress would be slow, small control inputs with a lot of compensation. This presents a three dimensional problem for our build-up with stress being defined by the compensation, rate, and magnitude of control motion. This build-up in stress is illustrated in Figure 11. The obvious problem is how to build-down in compensation? There is simply no easy answer on how to build-down in compensation as most compensation is subconscious and/or precognitive and something we just do without thinking about. It would be even harder to measure as the timing of inputs becomes critical to compensation and there is no way to measure that to see if the technique is being done properly.

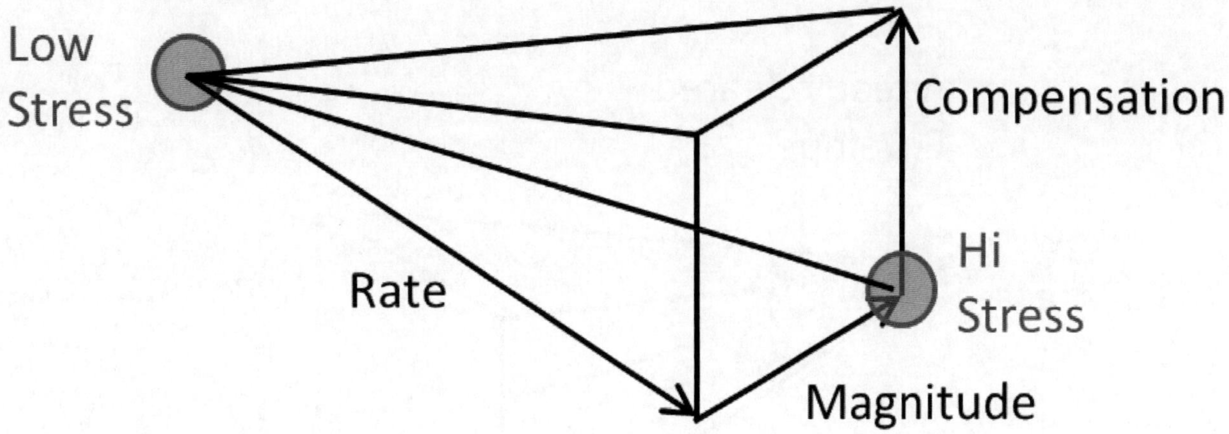

Figure 12, Modified Piloting Universe

Fortunately there is no need to know how to build-down in compensation or even define what it is or how to measure it. The piloting universe is not actually a cube as described in Figure 11. Let's consider the zero rate plane of the envelope. With zero rate, there is no motion and without motion there can be no compensation and no magnitude. The envelope is, in fact, a pyramid as shown in Figure 12.

As you can see in Figure 12, a build-up in stress from the low rate/low magnitude corner towards the high rate/high magnitude/zero compensation corner can be done without ever compensating. In other words, compensation is irrelevant to a HQ Stress test build up, a distraction at best and a barrier to a smooth buildup in stress. Since the aircraft managed to be flown to the point of beginning the FTT, there is a small corner of the HQ Stress envelope that will have already been explored as shown in Figure 13. An actual HQ Stress test build-up is illustrated in Figure 13 from the red point to the end point staying on the zero-compensation plane the entire time.

The idealized HQ Stress Test FTT would then begin with zero compensation and build up in rapidity and magnitude of control input. The size of the control inputs are controlled and independent of the error signal being tracked. It is closed loop tracking where the input rate and magnitude are controlled and the resulting track error is unimportant. For that reason the technique could accurately be called "Controlled Amplitude, Rate Build-up, Zero Compensation Tracking" (CARBZC Tracking). Those familiar with the SIS PIO technique can think of this as a modified SIS PIO with rate build up or "rate Buildup SIS PIO". One could also think of this as a closed loop rate and magnitude sweep. A frequency sweep is an open loop maneuver, so do not confuse a closed loop sweep with the classic open loop frequency sweep. Since we are no longer concerned with compensation, our HQ Stress test becomes a two dimensional problem of inceptor rate and inceptor magnitude (Figure 14). Small and slow inputs with zero compensation would be at the low HQ Stress end of the spectrum. We then increase the rate and size of our zero compensation tracking until we reach an aircraft limit or we reach the point where the stick is being moved as fast and as far as we are capable of.

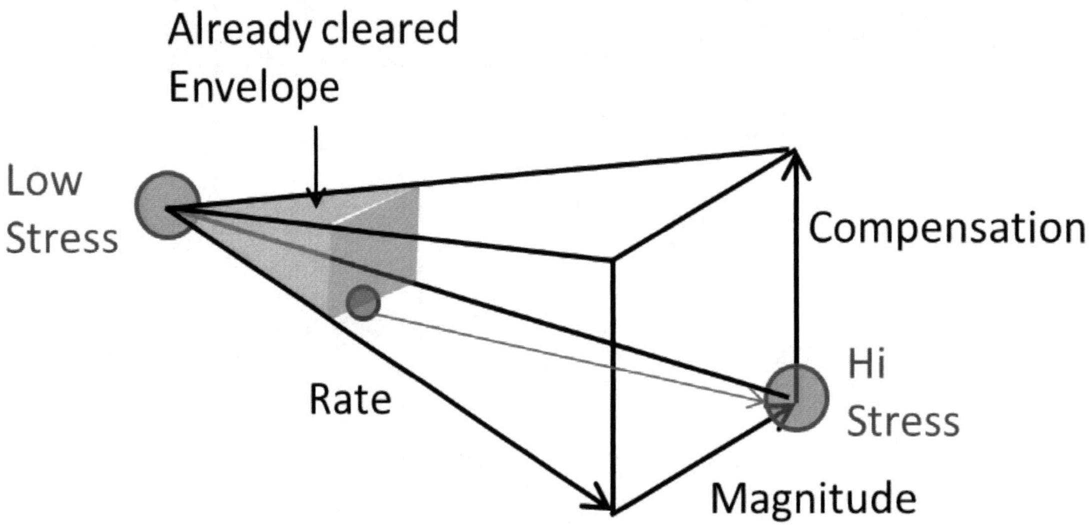

Figure 13, HQ Stress Test Build-up

We could, theoretically build up in both rate and magnitude but most pilots would probably find that difficult. Instead, a pilot could build up in rate first while limiting the size of their inputs to some preset small value. As the pilot builds up in rate, eventually they will be moving the stick (inceptor) very quickly but no larger in size than the original input. This point would be a small amplitude synchronous control strategy. If the aircraft behaves satisfactorily, a build-up in magnitude as illustrated in Figure 15 can be performed finally reaching the upper right corner of full amplitude bang-bang control. This last portion of the build-up is identical to that used at the end of the HQDT technique being used by TPS in 1999. It was a high stress way to finish clearing the envelope that the original HQDT technique began but did not fully clear.

Since this technique is always with zero compensation, there can be no performance goal associated with it. Good track performance will always require compensation so a zero compensation technique will never achieve optimum track performance. Tracking performance is not a goal and compensating to make it better destroys the build-up in stress. For that reason, there can be no performance criteria or boundary avoidance criteria associated with a HQ Stress Test. Cooper-Harper ratings, which rely on a relationship between performance and workload, are not applicable to this type of build-up.

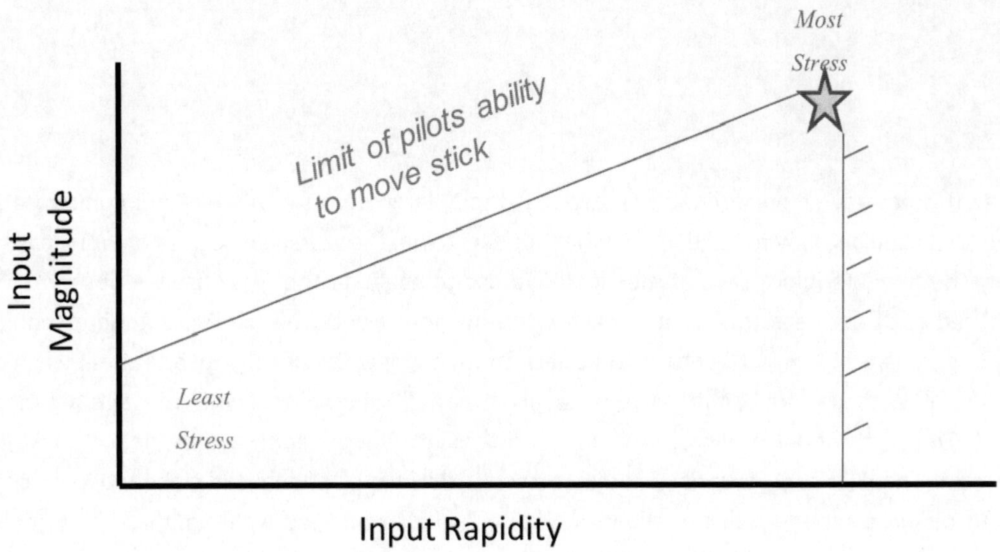

Figure 14, CARBZC Build-up Envelope

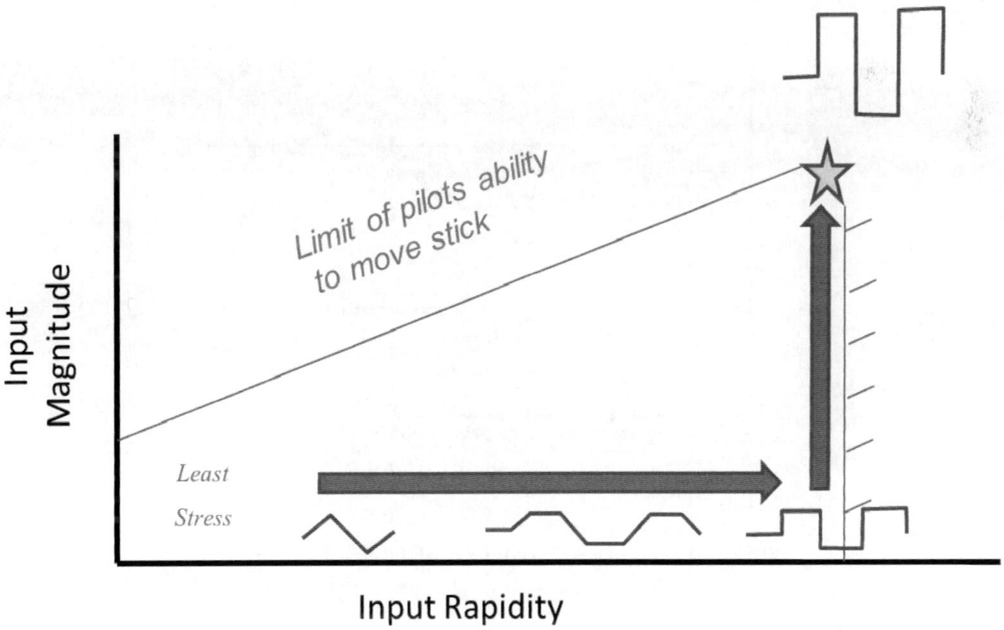

Figure 15, Recommended CARBZC Build up

The Metric

At this point we have defined HQ Stress as rapid, large inputs made without compensation. We have also identified one way to build up to, what appears to be, the worst case stress or worst case piloting behavior. The final roadblock is what metric do we use to evaluate the response we observe? We have already ruled out track performance as a metric so how do we determine if it is a good response vice a bad one? For this problem, let's return to Rogers Smith's work. Specifically inspirational was the Neal-Smith criteria. With the Neil-Smith criteria, a pilot model is combined with the aircraft model and evaluated. The pilot model or "Pitch Attitude Compensator" has a gain, a time delay, and a lead/lag compensator. The pilot compensation is adjusted so as to achieve an idealized closed-loop response and the extent of the compensation is evaluated. Figure 16 illustrates the Neal-Smith criteria goals. Good closed-loop behavior is defined as a small low frequency droop and a minimal resonant peak requiring little pilot compensation. In other words, we want consistent response up to some critical bandwidth[8] (ω_{BW} in Figure 16) with minimal pilot compensation. Ideally, we'd like consistent response with zero compensation. The critical bandwidth is a function of aircraft mission with some mission elements demanding a higher critical bandwidth than less demanding tasks.

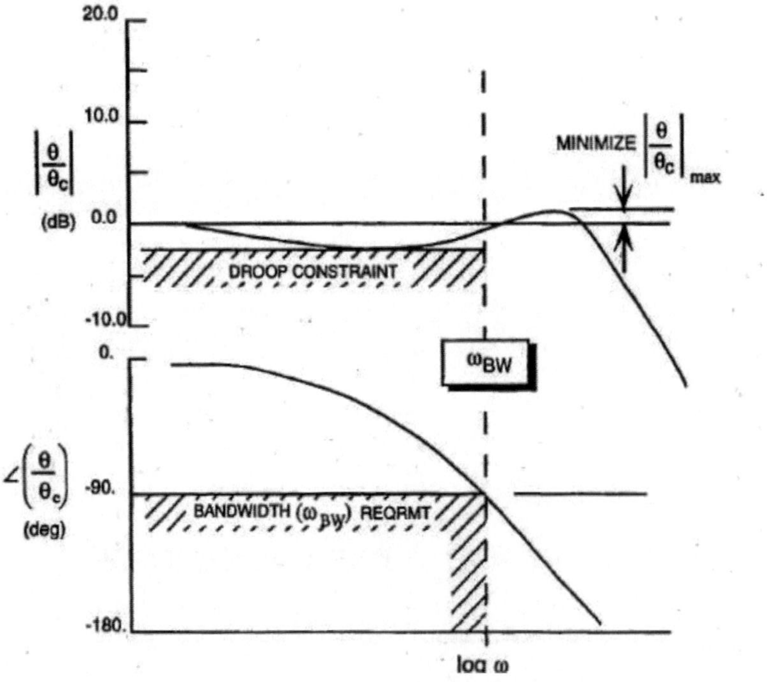

Figure 16, Neal-Smith Tracking Performance Standard

[8] Neal-Smith calls this parameter the "closed-loop bandwidth" which is defined as the frequency at which the closed-loop phase is -90degrees.

Since the CARBZC technique uses zero compensation throughout, an inverse Neal-Smith criteria seems appropriate. If the Neil Smith Criteria asks "How much pilot compensation is necessary to achieve an ideal closed loop response?" The Inverse Neil-Smith Criteria would ask "Given zero compensation, how close to an idealized response is the closed loop behavior?" The Idealized response in the Neil-Smith criteria essentially describes a response that is consistent and without delay up to a frequency necessary to accomplish the mission (ω_{BW}). So that is our metric, predictable response without delay up to a critical bandwidth and then a predictable decay in gain and increase in phase delay would be acceptable. Notice that this is like Rainbow Canyon in AutoGCAS. We have not defined the critical bandwidth yet. An aircraft might exhibit a behavior during the CARBZC build-up that would be suitable for one mission but that same response would be unsuitable for an aircraft that has a more demanding mission. The CARBZC metric is not a rating, it is only a way to identify potential HQ deficiencies and phase lag is a potential deficiency if it occurs at frequency necessary for mission accomplishment.

As we increase inceptor rapidity and then inceptor magnitude from low stress to maximum stress, do we get surprised? Are there any abrupt changes in aircraft response? It is important to point out that the definition of predictable is made in the context of the input made. The question is: "did I get the response I expected, given the input I made?" The question is not: "did I get a desirable response that reduced tracking error?" Performance is neither a metric nor an objective. It is critical that the evaluation pilot not use performance as a metric or to strive to achieve some level of performance. This is not an easy task as humans intuitively try to track error signals. Compensatory tracking behavior is a natural behavior and if we set track performance as a goal, we will modify our control strategy to achieve that goal. By allowing track error to drive our inputs, we defeat the build-up. If we build-up in stress slowly, we have an opportunity to recognize a deficiency and recover prior to experiencing the ill effects of that deficiency. When a pilot tracks for performance, control inputs are driven by the track error and not by the desire to follow a controlled build-up. If the target makes an abrupt change or the track error builds quickly, the pilot striving for performance will make an abrupt correction. In essence, the track error can be a trigger that will drive the pilot abruptly to high stress. The CARBZC technique will not yield optimum tracking performance, but that is not its purpose. The use of Cooper-Harper ratings, or any other performance based metric is incompatible with this technique. This HQ Stress test intentionally does not compensate yet some compensation will always be required to optimize track performance. Predictable response is the appropriate metric not track performance.

While observing the entire build up in stress, the pilot's input size should control the steady state response and the input rate should control the initial aircraft rate. Likewise, the size of input should ONLY determine the size of the response and should not have any effect on initial rate, or phase lag. The same is true for input rate. An increase in rate of application of an input should not affect the steady state response. For example, if a pilot puts a 10lb input in slowly and gets a 3g response, that pilot would expect a 3g response even if he was scared and put that 10lb input in very quickly. Simply said, rate controls rate only, size controls size only.

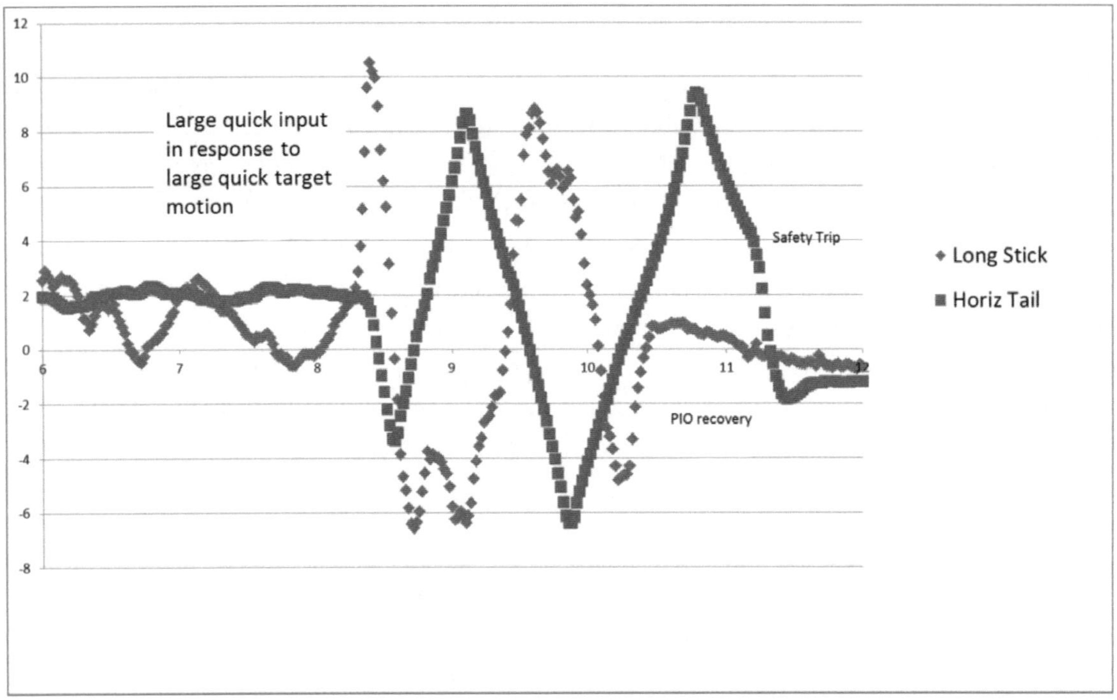

Figure 17 Operational High Gain Tracking on Rate Limited Aircraft

Predictable Response Metric Example

In order to illustrate the concept of predictable response as a metric, a short demonstration was performed on the Variable Stability In-flight Simulator and Test Aircraft (VISTA) with the results shown in Figure 17 and 18. The VISTA was programmed as an unstable aircraft that used stabilator control power to augment the stability much like an F-16 would. That stabilator actuator was programmed with a rate limit which resulted in a handling qualities cliff. During operational tracking evaluations, the jet routinely departed when the rate limit was inadvertently encountered. At some point in the operational tracking evaluation, the evaluation pilot would correct for a sudden large error and that would trigger the rate limiting and, because the bare airframe is unstable, quickly results in a departure. During these evaluations, the deficiency presents itself as a cliff with no apparent warning. Figure 17 shows the results from an operational tracking evaluation using high gain. At approximately 8.5 seconds into the evaluation, the target maneuvers quickly and the sharp correcting input by the pilot (blue trace) triggers the rate limiting and a divergent pitch oscillation results after about one cycle of PIO. Figure 18 shows the same aircraft while performing the latter half of the CARBZC technique. In this case, the CARBZC technique is being executed by a closed loop PTI (blue trace in the figure). The computer will close the loop with zero compensation and nearly zero time delay. The magnitude of the PTI input is increased incrementally mimicking a pilot performing the latter half of the CARBZC technique. If the aircraft was "predictable" the size and frequency of response should be consistent with the input. With the rate limit deficiency, this is not the case. Initially, the frequency of closed loop oscillations is 2.5hz and a gain of 0.4. Here, the term "gain" refers to the size of the output in degrees/second versus input signal in pounds of stick force and does not refer to "pilot gain". When the size of the pilot's input is increased, the frequency decreases to 1.7hz and the gain increases to 0.58. In other words, the response gets slower and bigger. We expect a bigger response with a bigger input but in this case, the size grows disproportionately with the increased input. This is seen as a non-linear behavior by the evaluation pilot. Also, we would not normally expect the size of our input to have any effect on the speed of response. As input size increases further, the frequency drops to 0.84hz and the gain is more than double the original at 0.84. When the size of the pilot input is increased once more, the resulting response begins to diverge, with a safety trip on the sixth overshoot. There are two obvious cues of an impending problem. First, the response is magnitude dependent. A change in input size does not yield a correspondingly proportional increase in response size. Further, there is a change in response frequency. Finally, even with a constant size input, the cycle to cycle response changes. In this case, two identical inputs result in different aircraft responses. This presents clear warning signs that something is amiss well before the actual departure. Contrast this with the operational tracking task where the deficiency came as a surprise "cliff".

Compare the results shown in Figure 18 and in Figure 19. In Figure 19, the maneuver was performed on an aircraft without a cliff and without significant deficiencies. In this case, you will notice there are no noticeable changes in gain or closed loop frequency as the stress is increased. In other words, regardless of how stressed the pilot is, the response will be proportional to the input and consistent across the envelope. In these two examples, only the latter half of the CRABZC technique was performed and it was implemented by a PTI. The PTI was used only for illustration purposes. By strictly controlling the input size and compensation, it is easier to illustrate the response differences. The technique is not intended to be flown automatically nor required to be flown automatically to achieve good results.

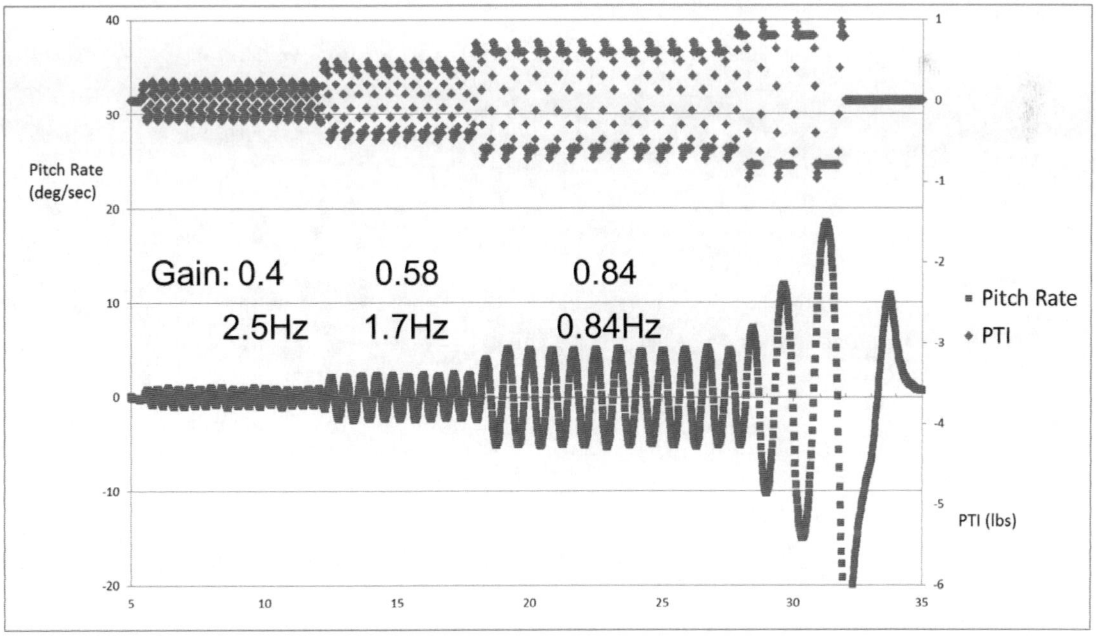

Figure 18 Pitch Rate CARBZC Flown with PTI - Rate Limited Aircraft

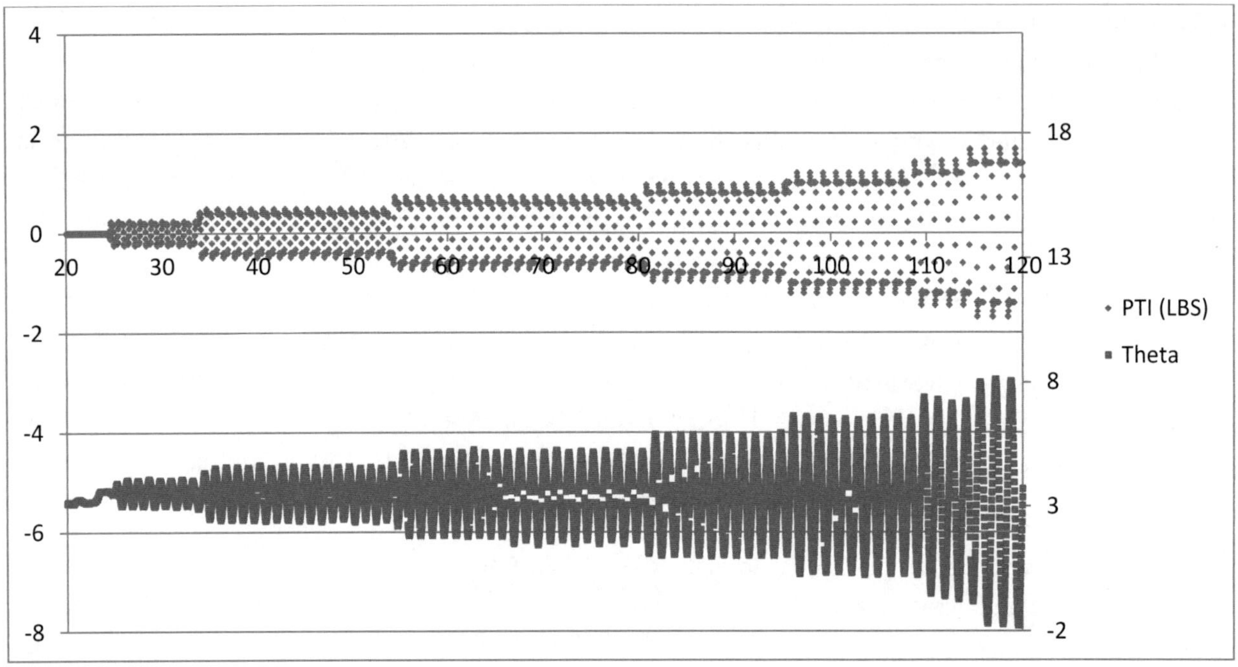

Figure 19, Pitch Rate CARBZC Flown with PTI - Level 1 Aircraft

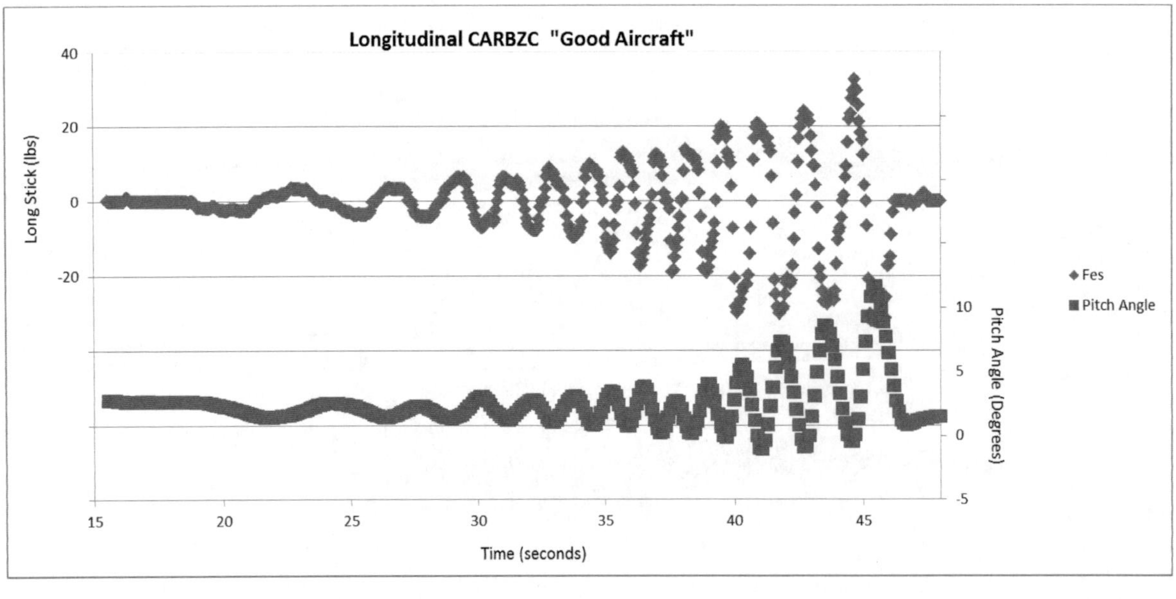

Figure 20, Manually Flown CARBZC in Level 1 Aircraft

Figure 20 illustrates the results of the maneuver being manually flown in an F-16. The F-16 does not have any HQ cliffs nor does it have any significant deficiencies in the cruise configuration. In Figure 20 you can see that the response gain and closed loop frequency does not appear to be influenced by input stress. The aircraft response follows from the pilot inputs and would be seen as a predictable aircraft.

Flight Test

Some experimenting was done from 2013 through 2015 to evaluate the potential of this technique in the real world. However, the flight testing was neither robust nor was it scientific. It was performed in conjunction with student training sorties and the students had very little HQ stress test training. The students also had no training, prior to the flights, in some of the techniques being evaluated to include the CARBZC technique. The results were promising so they are presented now only to inspire more formal testing.

Test Item

The VISTA (Figure 21) was used for this test and programmed with four aircraft configurations. The first was a good (but not perfect) aircraft. The second was a level 3 aircraft due to a very lightly damped short period mode. The third and fourth were rate limit configurations. One of these was an F-16 like configuration that was unstable in pitch and relied on active control of symmetric stabilators to maintain stability and those control surfaces were rate limited. This was identical to the configuration used in the example discussed previously. The other rate limited configuration had rate-limited elevons used for pitch control and pitch stability as well as for roll control.

Figure 21, VISTA NF-16D[9]

The subjects of the test were various students that were given a handling qualities demonstration sortie as part of the TPS curriculum. The demonstration was designed to introduce the student to the use of BAT WLB on a mystery aircraft with unknown latent HQ defects. This sortie was used to collect data by giving the students the opportunity to fly any HQ Stress test FTT on any configuration they chose. The

[9] Photo courtesy USAF TPS

students had no formal training in the CARBCZ technique, KLONK method, or the HQDT technique. They had some training on BAT and SIS PIO prior to this demonstration flight. Further, some of the students were not pilots. Only the CARBZC data from manually flown maneuvers were used in this study, the automated maneuvers were demonstrated to some students but their evaluation of the results was not used here.

The evaluation criteria

The evaluation used a three point scale to keep things manageable in the cockpit during a busy student training sortie. The three point scale was as follows. The best rating would go to the technique that identified the deficiency and did so safely. Safely is defined as no Variable Stability System Safety Trip. The Safety Trips in the VISTA are designed to, among other things, prevent departures and over-gs. The middle rating would be given to the technique that identified the deficiency but did not do so safely. Finally, the lowest rating went to the technique that never even found the deficiency at all.

Test Results

Initial results were quite promising for the CARBCZ technique. None of the students elected to try the HQDT technique but many applied a BAT technique in a way that was effectively identical to the way the HQDT technique was taught at TPS in 1999 but without the synchronous end point. The CARBCZ did outperform the KLONK, the SIS PIO, the BAT WLB/HQDT, and the operational task FTTs.

Level 1 aircraft:

For the level 1 aircraft, there was no particular intended deficiency although it was not programmed to be a perfect aircraft for all missions. The only purpose was to ensure that a particular technique did not suggest a serious deficiency when one did not exist. All the techniques worked well in that none identified a significant deficiency that did not exist. A couple students expressed difficulty with the CARBZC when phase and gain started to roll off beyond the ω_{BW} point. When you refer back to the Neal-Smith criteria, you will find a very interesting dilemma. What should the ω_{BW} be in the first place? The criteria suggests an answer that is mission dependent. In other words, you can't say whether the roll off was too soon or not unless you know the mission requirements. Ideally, a good HQ Stress test which goes to the endpoint of the HQ Stress spectrum should identify the roll off in phase at extremely high frequency regardless of mission. Mission suitability cannot and should not be determined strictly by the presence of this roll off since all airplanes exhibit it. It is only if that roll off in phase occurs at a frequency necessary to accomplish the mission that there is a problem. The CARBZC technique showed a roll off in gain and phase which could be a deficiency if it occurs at too low a frequency for the intended mission. You must then look to the mission requirements to see if that is the case. The CARBZC technique was the only technique that identified this roll off in gain and phase that occurs only at very high stress levels. This gives confidence that the CARBZC technique is truly spanning the entire HQ envelope to include the part that is higher "gain" than necessary for mission accomplishments. It also reinforces our original observation that a stress test does not replace an operational evaluation.

Level 3 Low Damping:

All the HQ Stress Test flight test techniques found this deficiency as it manifested during almost every maneuver. However, the SIS PIO and CLONK method sometimes had a safety trip. The safety trip keeps the VISTA safe when an unsafe pilot plus aircraft response is occurring. In these cases, the safety trip prevented an AoA overshoot that could have led to a departure or over-g. This makes sense because the SIS PIO and CLONK are essentially the endpoint of the CARBZC test and, if there is an HQ cliff, they will find it by falling off of it.

Rate Limiting Configurations:

This is where the CARBZC really shined. This configuration would exhibit no deficiency at all until the rate limit was approached. Once rate limiting occurred, a departure would immediately ensue. The operational task was the least effective here. It either failed to find the deficiency or the pilots found it while they departed controlled flight. In the context of this task, the warning signs were too subtle to be noticed or the pilots simply stepped over them with an abrupt increase in aggressiveness (stress). If we plot inceptor rate and magnitude over time for an operational task, the rate and magnitude change dramatically with time even if there is a build-up in aggressiveness (or gain) as shown in Figure 23. In Figure 23, you can see the rate (black line) and magnitude (red line) of pilot inputs are constantly changing and abruptly go high stress at times. Even when pilot gain is low, some rapid and large inputs will be used to achieve the desired track performance. A build-up in gain simply does not allow for a controlled build up in stress. The result is that a deficiency like rate limiting will manifest as a HQ cliff. This demonstrates why no performance criteria can be used with HQ Stress testing. Any performance goal baits the pilot into compensating and any compensation will destroy the controlled stress build up and potentially take the pilot over the cliff if it exists.

The SIS PIO and CLONK found this deficiency almost every time but also "departed"[10] almost every time. The BAT WLB found it often but not as consistently as the CLONK, SIS PIO, or CARBZC. The reason I believe this happened was that, as the boundaries shrank, the pilot compensation increased and with that compensation came quicker but smaller inputs. Since rate limiting is sensitive to input size, the technique frequently failed to identify a deficiency because the pilot's inputs tended to get small as they got more aggressive. Also, the BAT WLB would depart occasionally when it did find the deficiency. I believe this was because the pilot aggressiveness could sometimes jump if they got near a boundary early in the evaluation. In this case, there

[10] In variable stability aircraft, the model can depart but the Safety Trips will prevent the host aircraft (VISTA) from actually departing controlled flight. The departures noted here were safely protected by the host aircraft Safety Trips and there was no real danger

appears to be two different behaviors to boundaries. When the boundaries were in view and an adequate margin to them existed, the pilots used small, quick, high compensation inputs to keep clear. However, if ever surprised by a boundary, the pilots responded with rapid, large, reflexive inputs (high stress inputs) to avoid it. A distinction between known boundaries and surprise boundaries existed. With known boundaries, the pilots built up identically to a gain build up like HQDT. Surprise boundaries created a trigger where the inputs abruptly became high stress and Klonk like. Often this increased stress would jump right past the warning signs and a departure would result.

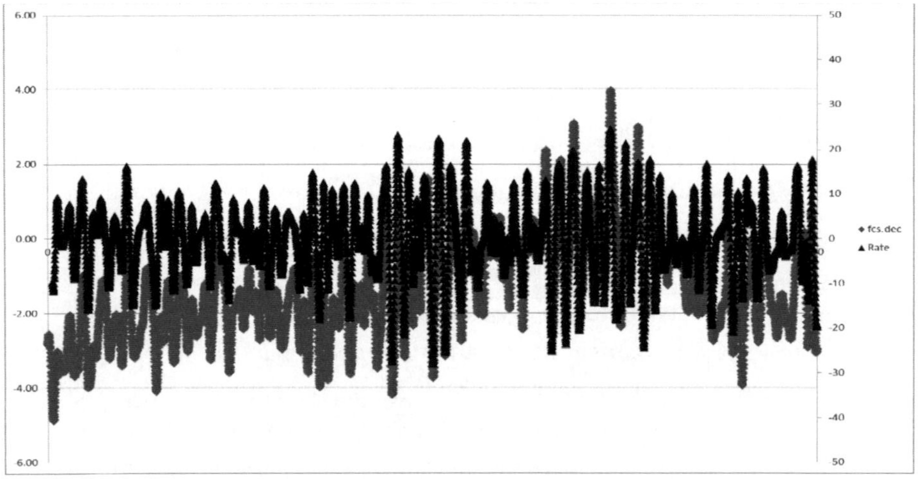

Figure 23, High Gain Tracking Stick rate and magnitude over time

The CARBZC worked exceptionally well with this configuration however. The deficiency was identified every time the technique was properly flown and it was very rare to see it result in a departure. When it did depart there were clear warning signs that were simply ignored by the students as they got wrapped up in the task. Often when this occurred the students instinctively reverted to a compensatory tracking methodology in an attempt to correct a sudden error induced by the target maneuvers. This sudden reversion to compensatory tracking would defeat the buildup and plunge the student off the cliff. The reversion to compensatory tracking was not uncommon and the tendency was strong with some subjects having a lot of difficulty with the concept of closed loop tracking with no compensation. Some of these students were given a brief critique and allowed to repeat the maneuver and, on subsequent attempts, were usually (but not always) able to find the deficiency while avoiding a departure. In fact, with some students, we made a game out of it where the students would move back and forth on the HQ Stress scale to incite the deficiency then back away so they could once again poke it. In a very short time, these students were able to consistently identify the incipient phase of the departure without departing controlled flight. The technique did require training however. Tracking to minimize errors seems to be the natural and preferred method of tracking a target and zero compensation tracking that allows errors to build seems to be unnatural. The obvious conclusion is that this technique does require training and proficiency to some degree. At the start of the data collection, a moving target was used for all the techniques except Klonk. It became apparent that the moving target was baiting some of the pilots into compensatory tracking during the CARBZC evaluation. A moving target is traditionally used to help pilots increase their "gain". However, for zero compensations techniques like the Klonk, CARBZC and SIS PIO, a moving target is totally inappropriate. When the target was kept fixed, the number of subjects that reverted to compensatory control dropped dramatically and consistency improved tremendously. The obvious lesson is to use a task appropriate to the technique and the test goals. Any task designed to increase pilot compensation is not appropriate for a test that relies on zero compensation.

When the CRABZC technique was properly performed, the "predictability" Inverse Neil-Smith metric proved effective in cueing the students into an impending problem. As soon as they got a wildly unexpected result, they could easily prevent the departure by getting out of the loop.

Conclusions

Admittedly the flight test was not scientific but the CARBZC technique did earn a shot at additional more formal research and discussions. More importantly, however, we should all challenge our assumptions as to what "gain" is and what "stress" is. We should also consider build-downs in compensation vice build-up when looking for HQ deficiencies.

REFERENCES

Smith, (June 1977), "A Theory for Longitudinal Short Period Pilot Induced Oscillations" AFFDL-TR-77-57

Neal, Smith, (December 1970), "An In-flight Investigation to Develop Control System Design Criteria for Fighter Airplanes" AFFDL-TR-70-74

McRuer, "Pilot-Induced Oscillations and Human Dynamics Behavior" NASA Contractor Report 4683

Witte, "An Investigation Relating Longitudinal Pilot-Induced-Oscillation Tendency Rating to Describing Function Predictions for Rate Limited Actuators" AFIT/GAE/ENY/04-M16

Committee on the Effects of Aircraft-Pilot Coupling on Flight Safety, "Aviation Safety and Pilot Control, Understanding and Preventing Unfavorable Pilot-Vehicle Interactions", National Research Council Washington D.C. 1997, Available online at: http://books.nap.edu/catalog.php?record_id=5469

Smith, (11 Nov 2008), "Handling Qualities Review...a Practical Perspective", http://www.dglr.de/fileadmin/inhalte/dglr/fb/l6/l62/2008-11-11_symposium/Programm_Vortraege/02-R_Smith_DGLR_Manching_081111.pdf

Dotter, "An Analysis of Aircraft Handling Quality Data Obtained From Boundary Avoidance Tracking Flight Test Techniques - Thesis", AFIT/GAE/ENY/07-M24, Air Force Institute of Technology, Wright-Patterson AFB, Ohio

Dakoogh & Berefelt, "Clearance of Flight Control Laws using Optimisation, AST5-CT-2006-030768"

Niewind, (2011), "What the Hell is Pilot Gain?", German Aerospace Center, Institute of Flight Systems, Flight Test Branch

Prosser, Thurling,(2000), "Handling Qualities Stress Testing", USAF Test Pilot School. Available online at www.setp.org

Lee, "Recent Experience in Flight Testing for Pilot Induced Oscillations (PIO) on Transport Aircraft", Boeing Commercial Airplanes Seattle, WA

Mitchel, "Development of Methods and devices to Predict and Prevent Pilot-Induced Oscillations" AFRL-VA-WP-TR-2000-3046, Hoh Aeronautics Lomita CA

Gray, "A Boundary Avoidance Tracking Flight Test Technique for Performance and Workload Assessment", USAF Test Pilot School

Angner, Jensen, and Seidl ,"JAS 39 Gripen EFCS How to Deal with Rate Limiting", Available online at www.setp.org

Dennision, McMeeking, Florendo, Haywas, and Newton , (June 2014), "Limited Investigation of Aircraft Response Using the Switch-Induced Simulated Pilot Induced Oscillation Flight Test Technique (Have Thrash), USAF TPS-TIM-13B-03

Flying Qualities of Piloted Aircraft, MIL-HDBK-1797, Department of Defense Washington DC 20402. Note: MIL-HDBK-1797 is a reprint of MIL-STD-1797A redesignated as a handbook.

Celere, Varoto, and Maciel, (2007), "Verifying Pilot Gain in PIO Flight Test", http://arc.aiaa.org/doi/abs/10.2514/6.2007-6640

X-2 Starblaster

At first glance the Bell Aircraft Corporation of Buffalo, New York may have seemed a strange choice to become a hotbed of leading edge experimental flight testing in the post-World War II era. After all, Bell was not really one of the big established airframe companies such as North American, Boeing, or Grumman, and its war effort had not produced the fastest or highest flying operational fighter aircraft of World War II – just the peculiarly configured P-39 and P-63 mid-engine designs shipped by the thousands to the Russians to later disappear in the frigid battlefields of the war's Eastern Front.

So what made Bell a good choice for the world's most advanced high speed flight research aircraft? For one thing, they had successfully developed the United States' first jet powered aircraft with the P-59 Airacomet. Even though only 68 or so of this aircraft type were built and the jet's performance fell short of expectations, the design and flight testing experience established Bell as a forward thinking aircraft manufacturer who was not afraid of new and novel technology. Bell also offered a convenient surplus of available production line space supported by an in-house stable of experienced engineers and designers with an aggressive attitude towards research and development. These ingredients proved to be enough to elicit a handshake agreement between Bell's Robert Woods and NACA's Ezra Kotcher on November 30th, 1944 to begin work on an aircraft specifically aimed at breaking the sound barrier; the XS-1[1].

Built and tested under a joint NACA and US Air Force contract, the Bell XS-1 (No. 46-062), represented a substantial departure from conventional aircraft design with a bullet shaped fuselage, an extremely thin (only 8% chord thickness[2]) straight wing, liquid fueled rocket propulsion, and only one small entrance door on the right side of the fuselage offering little chance of pilot escape. The airplane was a point design solely focused on defeating the "Sound Barrier" with all of the Bell engineers' hopes pinned on their calculations that the 6,000 pound thrust Reaction Motors XLR-11 rocket engine coupled with the XS-1's thin wing and bullet shaped fuselage could do the job at hand. Those hopes and calculations were proven correct when the airplane, now simply called the X-1 and named "Glamorous Glennis" after pilot Chuck Yeager's wife, was carried aloft by a B-29 "mother ship" on 14 October 1947 and became the first aircraft to exceed the speed of sound by hitting 1.06 Mach (700 mph) over the Muroc Dry Lake bed at what later became Edwards Air Force Base[3].

With the X-1 and Yeager's successful challenge and conquest of the sound barrier demon, Bell found itself in the position of not only having built the world's fastest aircraft but also having become the US Government's preferred provider of rocket powered research planes. During the late forties and early fifties, Bell produced another two X-1 aircraft (No. 46-063 and 46-064) and a succession of three later X-1 variants; the X-1A, X-1D, and X-1E. A total of six different X-1 airframes flew with varying degrees of success up into 1958[4] ultimately achieving a top speed of Mach 2.44. Diverging from the X-1 line, Bell also constructed a pair of XS-2 Starbusters (later shortened to just "X-2" like the earlier X-1) powered by newer design, higher thrust rocket motors built at the nearby Curtis-Wright Company in Caldwell, New Jersey.

The Starbuster's basic design features were laid out in preliminary engineering meetings chaired by Stan Smith in 1945 after he was assigned lead engineer duties for the company's MX-653 program on May 11th of that year[5]. These meetings set the basic features of the MX-653, soon to be known as the X-2, including the decision to air-launch by mother ship and to incorporate a 40 degree wing sweep on both the main wing and horizontal tail. It was also decided to build both the main wings and horizontal stabilizer from stainless steel which, coupled with the craft's tiny frontal area and exotic K-Monel nickel alloy fuselage, was specifically intended to improve upon the performance of the straight wing X-1 series research aircraft and permit the X-2 to investigate aerodynamic heating phenomena while striving for even higher and faster goals; namely Mach 3 and the yet-to-be-attained 100,000 foot altitude mark[6].

One of the X-2's most striking features was its small size and extraordinarily confining cockpit. In order to provide the least possible amount of frontal area, Bell designers crafted the crew compartment to suit small pilots only with minimal headroom and limited adjustability; after all, the flights would be short and always conducted in good weather so crew comfort (and fit) were very low priorities. The cockpit and the X-2's low profile, close fitting canopy were so tight that pilot Pete Everest described them thusly in a later interview, "The tiny X-2 cockpit was so small that your helmet was touching the canopy on both sides, while your legs and feet were straight out in front of you on the rudder bars; your shoulders touched the canopy rails on both sides."[7] Certainly not the description of a cockpit suitable for a tall man, let alone anyone that might be the least bit overweight or suffer from a touch of claustrophobia!

The small size of the rocket plane mandated the use of special smaller diameter gauges for all non-critical flight instruments[8], preserving the precious panel space needed by full size gages for only the most crucial information such as airspeed, altitude, and a single artificial horizon. In a nod towards aircrew safety, the X-2 did incorporate a novel pilot escape system that separated the entire forward fuselage section in the event of a high speed mishap. The pilot was then expected to manually bail out once the cockpit capsule reached a lower speed and altitude.

The X-2's diminutive size was also evident in the 40 degree sweptback wings' small 32 foot span and an empty weight of slightly less than 12,500 pounds. Once hoisted into position below its mother ship, fueled with oxidizer and rocket propellant, and readied for flight, the aircraft weighed approximately 25,000 pounds and could depend upon as much as 15,000 pounds of thrust from its Curtis-Wright XLR-25 two-chamber rocket motor: a 2:3 thrust to weight ratio that would rapidly approach and surpass 1:1 as the fuel and liquid oxygen burned down while feeding the hungry motor.

A unique and innovative feature at the time was the ability to infinitely vary the XLR-25 rocket motor's thrust from approximately 2500 pounds right up to the maximum output of 15,000 pounds. Based on a Robert Goddard design developed under US Navy sponsorship at Annapolis during the closing stages of World War II[9], the motor ran on diluted ethyl alcohol fuel and liquid oxygen and was mounted in the X-2's aft fuselage with the two exhaust nozzles stacked one atop of the other in an effort to minimize thrust effects on the aircraft's pitch and directional stability. The motor in the first airframe, 46-674, was also mounted slightly off-axis by approximately 0.7 inches, producing a small asymmetric thrust that almost always resulted in the X-2 rolling left to somewhere between 20 and 30 degrees angle-of-bank during a full rocket burn[10]. This unique but predictable roll was just one more thing to deal with when flying this point designed aircraft.

The X-2 originally was to incorporate an advanced electrically driven Bendix Corporation flight control system powered by a series of fragile dry cell batteries[11], a system that turned out to be too far ahead of its time to cope with the demands of the high speed airplane. The Bendix system was eventually abandoned after consuming two years of effort and an estimated $4 million dollars' worth of investment. It was replaced by a "quick fix" purely mechanical flight control system for a short series of unpowered glide flights intended to amass some initial flying qualities data for the swept wing aircraft. Later on, the mechanical cable actuated rudder was retained (which was locked in place for supersonic flight) and an irreversible simplex hydro-mechanical set up installed to drive the all-moving horizontal stabilizer and ailerons for the remainder of the test program and for all powered flights. The resulting flight control system was a far cry from the advanced Bendix electrical system originally proposed and its use sacrificed any thought of stability augmentation or adaptive controls that may have been possible with the original design concept. While the hydro-mechanical system did offer more proven reliability and relative simplicity, it was far behind the state-of-the-art ideal when it was finally installed in the X-2 and left the pilots entirely on their own to deal with changes in flight control effectiveness as the X-2's stability decreased at higher speeds and the thin air of the stratosphere offered less and less control power at high altitudes.

In addition to delays caused by the complex flight control system, production difficulties attributed to welding of the K-monel and stainless steel airframe structures contributed to the problems, with shortfalls in skilled labor and the massive nature of the welding equipment (it was advertised as "portable" by the manufacturer but reportedly took a two-ton crane to be moved[12]) blamed for the extra time and expense. On top of these delays, the Curtis-Wright XLR-25 rocket engine was not yet finished by the summer of 1952 when the X-2 program opted to install a highly simplified "quick fix" flight control system in the second airframe, No. 46-675, and use that system to conduct a series of unpowered flights to at least start collecting some preliminary flight data. The first of these glide flights took place on 27 June 1952 after a pair of earlier captive carriage flights had been carried out with the airplane firmly secured to a larger version of the venerable B-29 Superfortress, the B-50A; an aircraft otherwise best known for successfully completing the first non-stop around-the-world flight in 1952 . Two more glide flights and one collapsed nose gear later, the slightly bent X-2 was returned to the factory in New York for repairs as well as to receive its rocket motor and "production" hydraulic flight controls installation.

Finishing touches were also being applied to X-2 No. 46-674 but it wasn't until May 1953 that all systems and a working rocket motor were finally in place and flight testing of the X-2 was ready to begin in earnest. The second airplane, No. 46-675, was still ahead of its older sibling in the production process and was the first to be completed and tested. These tests took an immediate unexpected and tragic turn on May 12th of that year as on-board system checks were being conducted over Lake Ontario. A violent explosion occurred during testing of the liquid oxygen jettison system and both the X-2 and factory test pilot Skip Ziegler were tragically lost. The in-flight explosion took place while the research aircraft was still connected to its B-50A mother ship and resulted in not only the death of Ziegler but also that of Frank Wolko, a Bell technician onboard the host airplane. The B-50A mother ship was so severely damaged by the explosion that it was deemed unrepairable and scrapped after limping back to land at Niagara Falls while the explosion racked remains of X-2 No. 46-675 fell into the depths of Lake Ontario and were never found nor recovered.

Other research aircraft also inexplicably blew up during these early days of rocket plane testing including the third X-1, the X-1A, and the X-1D. Most of these explosions were so comprehensive that little remained of the rocket aircraft or their systems to aid in the subsequent mishap investigations while the lost X-2 carried any evidence of its causal factors to the bottom of Lake Ontario. After several false leads and incorrectly assumed causes, the true source of the rocket plane explosions was finally attributed to lowly leather gaskets used in the aircrafts' liquid oxygen plumbing. These gaskets, manufactured by the Ulmer Company of Norwich, Connecticut - a long time maker of leather products used in hydraulic and pump valve assemblies - used a tricresyl-phosphate substance in their leather treatment and tanning process to aid in waterproofing. Subsequent laboratory tests proved this chemical to be extremely shock sensitive and explosion prone after coming in contact with the -297 degree Fahrenheit liquid oxygen used in the rocket engines[13]. These seemingly innocuous gaskets directly caused the loss of four priceless research aircraft and serve as testimony to the fact that no detail is too small to cause a catastrophic mishap when it comes to experimental flight testing or rocket propulsion; a lesson tragically re-learned decades later with the loss of the space shuttle Challenger.

After the loss of the second X-2, Bell quickly went back to work on the original airframe, No. 46-674, to ready it for flight while a back-up B-50D was modified to replace the severely damaged B-50A mother ship. The research program then moved west to Edwards Air Force Base to begin an ambitious Air Force and NACA flight test program in the summer of 1954, with more glide flights commencing that June.

In addition to the planned acquisition of aircraft handling qualities data aimed at grasping a better understanding of swept wing aerodynamics, these non-powered flights also produced an unplanned series of hard landings, several of which necessitated a return to the Bell factory for extensive airframe repairs. In fact, the X-2 recorded at least five "landing incidents" in only twenty total flights, proving the low profile swept wing aircraft to be extraordinarily sensitive during the landing phase (if not in all phases) of flight and highlighting the risks involved in even routine aspects of this experimental flight test program. Modifications to the research aircraft's peculiar landing gear arrangement of a nose wheel, a main skid, and a pair of short stabilizing "whisker skids" mounted mid-span on the wings were made to lower the main skid height, increase the pressure exerted by the "whisker skids", and adjust the nose wheel caster[14]. These changes help cure the aircraft's iffy ground handling and propensity towards hard landings but still left the pilot with no brakes and only limited steering during the X-2's high speed lakebed landings and subsequent long slides/roll-outs to an eventual stop.

The airframe repairs and cross-country transportation logistics created by these landing anomalies further delayed the program, and powered flights of the X-2 were not achieved until 1956, when it finally looked as if the diminutive aircraft's issues were resolved and its touchy flying qualities understood well enough to press ahead. A full decade after its initial design work and with the November 30th, 1956 deadline looming for the aircraft to be turned over to NACA, Air Force speed and altitude record attempts were pushed to the forefront of the flight test schedule so the X-2 could claim its prizes before some other airframe beat it to the next aviation milestone.

Though Pete Everest flew the initial X-2 flights (including the glide flights of 1954), Air Force Captain Iven Kincheloe was selected to take over the project in the summer of 1956 and would serve as pilot for the altitude record flights with fellow Air Force Captain Mel Apt selected to fly the high speed runs.

Iven Carl Kincheloe Jr. was born in Detroit, Michigan on 2 July 1928 and graduated in 1949 from Purdue University with a degree in aeronautical engineering; just in time to enter the Air Force and serve as a jet fighter pilot in Korea from 1950 to 1951. "Kinch" proved to be quite gifted at his trade, reaching "double ace" status in Korea with ten aerial victories over communist MIG adversaries over the course of 131 combat missions flying both the F-86 and F-80 jet fighter aircraft[15].

Upon returning from overseas and completing a subsequent tour as a gunnery instructor at Nellis Air Force Base, Kincheloe's combat record, aeronautical engineering background, and solid reputation for precise and aggressive flying helped get him selected to attend the prestigious Empire Test Pilot School as a US Air Force exchange student. During his ten month course in England, Kinch split his days between academic lessons and flying while working with other British and foreign students to learn the intricacies of flight testing modern high performance aircraft. He graduated from Empire in December of 1954 and, to paraphrase the school's motto, now that Kincheloe had "learned how to test" he was ready to make his mark "testing to learn" and moved straight to the US Air Force's flight test establishment at Edwards Air Force base in 1955.

Once assigned to Edwards, Kincheloe flew a wide variety of chase and test aircraft including most of the century series fighters, which were just emerging from contractor developmental flight testing at that time. Gunnery testing of the F-100C soon led to more challenging flights in the brand new F-101 Voodoo, F-102 Delta Dart, F-104 Starfighter, and F-105 Thunderchief as their flying qualities, performance, and mission capabilities were measured and accessed prior to acceptance into the operational Air Force. Throughout 1955 and into 1956, Kincheloe's test sorties added up to a graduate education in flight testing through the hands-on school of practical experience and his reputation for airmanship grew along with his test acumen. When lead X-2 pilot Pete Everest was reassigned, it was an obvious choice for Edwards' management to move the dedicated and enthusiastic Kincheloe onto the program and begin to prepare for the rocket plane's assault on the record books before time ran out on the deadline to hand it over to NACA for their research flying.

By the time of Kincheloe's first flight in the X-2 in May 1956, the airplane had already been flown past 1900 mph and as high as 58,000 feet. Though the hard landings which had plagued early flights seemed to have been overcome, the X-2 continued to exhibit very relaxed stability in pitch and yaw at higher Mach numbers, creating concerns about controllability as speeds and altitude continued to increase. The degraded control effectiveness experienced in the thin atmosphere above 50,000 feet contributed to the aircraft's poor stability and X-2 pilots had to fly quite carefully and cautiously while zoom climbing away from the mother ship and accelerating to ever higher speeds. Pete Everest described a flight in the X-2 in the following manner, "Things happen so darned fast! It's just hanging on and trying to do the best you can. If you got up to an altitude where you wanted to 'push over' to get maximum speed, your altimeter was such that its indication was lagging behind, so you started the push over at about an indicated 5,000 feet below where your real altitude was, to compensate. You didn't want to push over too suddenly because you would un-port the tanks where they feed into the rocket engine. If you un-ported those and got air into the fuel system, the rockets would shut down. That occurred several times. So you flew very, very gently and did the best you could while flying like a bat out of hell!"[16]

The X-2's hybrid flight control system could not have helped the situation much, as the pilot had virtually no control over the aircraft's yaw axis once the mechanical rudder was "locked" into position for supersonic flight, and the gains built into the hydro-mechanical pitch and roll controls were, at best, just educated guesses at what would be needed to cope with the demands of exploring the uncharted high speeds and high altitudes made possible by the rocket plane. Coupled with the very tight fitting cockpit and the bulky yet restrictive X-1 type pressure suits that all X-2 pilots wore for their powered flights, the result was an aircraft that demanded close attention during every second of flight and was to be entrusted to less than a handful of highly experienced and gifted test pilots. In fact, including the aircraft's early glide profiles, only four pilots flew the X-2 during its brief 20 flight career and two of those died in the process.

Somewhat over two months elapsed between Kincheloe's first flight in the X-2 and his next, a practice run at high altitude which resulted in an 87,775 foot apogee on 3 August 1956 – high, but not high enough to establish a new record and almost three miles shy of the program's 100,000 foot goal. This flight did serve to familiarize Kincheloe with the demanding flight profile required in order to maximize the performance of the rocket powered craft and also allowed him to experience first-hand the degraded flying qualities of the X-2 as it approached ever higher Mach numbers in the thin air above California's Mojave Desert test areas.

In order to maximize the altitude gain from the X-2 it was necessary to first carry the rocket plane aloft nestled in the bomb bay of the B-50 mothership to as close to 30,000 feet as the overloaded piston engine bomber could attain, a process that might take as much as an hour after the pair's lumbering takeoff from Edwards Air Force Base. The X-2 would then be released at 250 mph and the rocket motor almost immediately started so as to quickly accelerate the compact research airplane while it climbed clear of the suddenly unburdened mother ship. The X-2 would soon outpace its chase aircraft and needed to be maintained at a very precise climb attitude so as to maximize the aircraft's performance and rapidly arrive at the planned altitude for the test flight. The time spent mired down in the transonic region where local shock waves on the airframe would contribute to a substantial drag rise would be minimized by gently unloading the normal G on the X-2, if not by diving it slightly to help accelerate through Mach 1. For an altitude attempt, a 45 degree climb profile would be aggressively captured after passing through Mach 1 and the rocket motor run at full thrust until the airplane's fuel and oxidizer supplies were exhausted, a process that usually took slightly over 2 minutes to accomplish.

One challenging aspect of this climb profile was that the natural horizon was located well below the nose of the aircraft and totally outside of the pilot's field of view throughout the entire climb and the single J-8 attitude gyro proved to be virtually useless for fine control of the airplane's pitch attitude. The close fitting cockpit and X-1 pressure suit's full coverage helmet limited the pilot's ability to turn his head more than a small amount from side-to-side, which eliminated the option of looking out the side of the aircraft and aligning the X-2's swept wingtip with the horizon to help gauge a 45 degree climb angle. Kincheloe and the Bell ground crews settled on a low technology solution in which a red grease pencil line was marked on the forward side of the canopy glass so that it was within the pilot's restricted field of view and could be used as a visual reference line when climbing. This proved to be a crude, but just effective enough, fix for X-2 pitch attitude reference and was used for the remainder of the powered flights.

After the rocket fuel ran out, the aircraft continued to trade speed for additional altitude as it climbed on a ballistic profile, with the final height depending upon the duration of the rocket burn, the aircraft's climb angle, and its maximum speed. At this point in the flight, the X-2 was really more of a projectile than a flying machine; with 98% of the Earth's atmosphere beneath the aircraft, flight control authority was, at best, quite marginal if not almost useless.

Coasting over the top of its arc, the X-2 would run out of excess energy and begin to fall back towards the Earth while the pilot waited for airspeed to build and the surrounding atmosphere to thicken to the point that gentle control inputs would begin to have a meaningful effect, first on the aircraft's pitch and roll rates and then, eventually, on its flight path. Smooth and steady were the key words for any pilot control movements, as the X-2 would consume thousands of feet of altitude before enough airspeed and atmosphere were on hand to establish adequate control effectiveness and the pilot could gently apply some positive G to initiate a shallow pull-out into a controllable glide profile. After slowing below supersonic speed and getting down to lower altitudes, it was simply a matter of visually finding the correct dry lakebed, coordinating a running rendezvous with the jet powered chase aircraft, and descending a fast and steep final glide to a landing on the expanse of the Roger's dry lakebed. Once the X-2's combination wing skid and nose wheel landing gear had been extended, the pilot had to touch down at more than 150 miles per hour and would then slide well down the lakebed before scraping to a halt trailing behind a cloud of dust and a small fleet of pursuing support vehicles. All this had to be accomplished while wearing a primitive X-1 pressure suit and helmet and strapped into a claustrophobically small cockpit, with limited chance of escape should something go wrong.

Kincheloe's third flight later in August did nothing to threaten the existing absolute altitude record. He was forced to shut down the XLR-25 rocket motor earlier than intended, as he could not maintain the requisite 45 degree climb profile and began to feel the lightly damped X-2 get more and more out of hand from a controllability standpoint. Ever mindful of the potentially catastrophic consequences that could accompany excitement of the roll-to-yaw inertia coupling tendency exhibited by the X-2 (which would likely lead to a total loss of control of the airplane), Kinch opted to try again another day and arced back down to the desert below to conclude the X-2's 18th flight.

That next opportunity came on 7 September 1956 when Iven Carl Kincheloe Jr. crawled down from the relative comfort of the B-50 crew compartment and strapped into the front of the X-2 suspended below in the bomb bay. Count down checks completed, he was released from the mother ship, ignited the X-2's rocket motor, precisely followed the prescribed climb profile, and topped out at 126,200 feet above sea level to destroy the existing altitude record and become, according to the national press, the "first of the spacemen"[17]

The X-2 had finally accomplished one of the goals it had been specifically designed for ten years before by becoming the first manned air vehicle to exceed 100,000 feet of altitude. Much to the US Air Force's pleasure, it had done so while being flown by a military pilot who also happened to be extremely well liked, extraordinarily well qualified, ready to speak to the press, and highly photogenic. According to Kincheloe's own words reported in *A Shadow on the Horizon*, "This height achievement was not only meaningful, in that I had over 99.6 [percent] of the earth's atmosphere below me and the concentration and nature of cosmic ray particles are the same as in actual space; therefore, space equivalent, but it is even more significant when one realizes the X-2 was a ten year old design."[18]

Kincheloe and the Bell X-2 Starbuster had done it; the first to fly above 100,000 feet and go up into the black sky where darkness was suspended above them while the Earth shone below. The pair had survived the almost non-existent atmosphere and low pressure of the stratosphere and successfully returned to a controlled landing back from where the flight had originated. For this feat, Iven Kincheloe was highly touted by both the press and aviation community alike and awarded a Distinguished Flying Cross and the Mackay trophy by the US Air Force for the "most meritorious flight of the preceding year". Unfortunately, nobody would ever match Kincheloe's altitude achievement in the X-2; the airplane would only accomplish one more record setting, but fatal, flight before its demise.

Now that the Air Force had surpassed 100,000 feet of altitude they made a final thrust toward adding Mach 3 flight to their list of accomplishments. Mel Apt was to be the pilot, in keeping with the original selection scheme, and a push was on to establish a Mach 3 speed record before to the Air Force would have to relinquish control of the X-2 to NACA, and before they lost the use of the dry lake bed to California's winter rains.

On 27 September 1956 the X-2 was carried aloft one final time for Apt's first and, tragically, only flight in a rocket powered airplane.

Apt had been a test pilot at Edwards since graduating from the US Air Force Test Pilot School in 1954 and had over 3,500 hours of flight time, 1,000 of which were in jet aircraft. Apt put his test pilot school and Air Force Institute of Technology aeronautical engineering educations to use flight testing many of the same aircraft as had Kincheloe as well as conducting inertia coupling tests of the F-100[19]. The X-2 assignment represented recognition of Apt's accomplishments and was a real opportunity to display his flying talents and move further into the realm of experimental flight testing and research. Though he had not yet flown the X-2, Apt had been part of the program for most of the year, had chased several of Kincheloe's flights, and had amassed many hours of cockpit time in the Goodyear Electronic Differential Analyzer[20] analog simulator practicing the demanding X-2 flight profiles[21]. When the time finally came to call upon his skills to take the little airplane faster than anyone had gone before, Apt was ready.

Re-topping the X-2's oxidizer and fuel tanks just prior to release from the B-50 mother ship, Apt managed the prescribed flight profile almost perfectly and squeezed an additional 12 ½ seconds of motor burn from the full-to-capacity tanks of the plane[22]. The result was a new world speed record of Mach 3.196 or 2094 mph[23]. However, Apt's extra speed and longer rocket burn had taken his flightpath well beyond the intended lakebed landing site and every two seconds carried him another mile further away from the Edwards dry lake bed. It is likely that Apt was concerned that the limited glide capability of the X-2 might fall short of a safe return, and this may have prompted him to apply too aggressive a control input so as to turn back towards Edwards before the streaking aircraft slowed enough to regain some stability. Whatever the reason, the X-2 violently departed controlled flight and entered what was later determined to be an inverted spin from which recovery could not be initiated.

Apt was probably in and out of consciousness as the airplane fell out of control but he did successfully activate the cockpit separation system and attempted to escape from the unrecoverable airplane as it fell through 40,000 feet. Unfortunately he was unable to jettison the canopy soon enough to jump clear of the cockpit section before it hit the ground and Mel Apt died within minutes of earning his title of the world's fastest man.

The X-2 program died along with Apt after only 20 flights, ten years, and millions of dollars of investment. It had gone the highest and fastest of all the early X aircraft, but by 1956 had also exhausted its capability to do much more. The aviation research world and pilot Iven Kincheloe were ready to move ahead to a new project destined to fly even faster and higher, the X-15.

Before Kincheloe could move on to the X-15 effort in earnest he continued to fly more routine test flights on Air Force aircraft, including the F-104 Starfighter. It was on one of these seemingly less demanding test sorties on 26 July 1958 that Kincheloe's F-104 suffered a low altitude engine failure shortly after takeoff from Edwards and he crashed into the desert, unable to safely initiate the stricken aircraft's downward acting ejection seat.

In recognition of his aviation accomplishments and dedication to experimental flight testing, the Society of Experimental Test Pilots (of which Kincheloe was an early member) named their most prestigious annual award the Iven Kincheloe Award. This award, presented every year since 1958 at the Society's annual symposium, recognizes the individual who, in the opinion of his or her professional peers, completed "the most outstanding professional accomplishment in the conduct of flight testing by a test pilot member of the Society of Experimental Test Pilots"[24]. The test flights must have been primarily conducted within the in the previous year and flown by the award recipient. This award is still revered today as the ultimate recognition available for the flight test profession and is a fitting tribute to the world's first "Spaceman".

[1] John D. Anderson, Jr., "Chapter 3: Research ing Supersonic Flight and the Breaking of the Sound Barrier" in *From Engineering Science To Big Science*, ed. Pamela E. Mack (Washington, D.C.: NASA History Office, 1998), page 86. Accessed February 2, 2016 at http://history.nasa.gov/SP-4219/Chapter3.html.

[2] "NASA - Dryden History - Historic Aircraft - X-1 Technical Data," NASA, accessed February 2, 2016, http://www.nasa.gov/centers/dryden/history/HistoricAircraft/X-1/techdata.html.

[3] "Bell X-1 Glamorous Glennis," Smithsonian National Air and Space Museum, accessed February 2, 2016, http://airandspace.si.edu/collections/artifact.cfm?object=nasm_A19510007000.

[4] "NASA - Dryden History - Historic Aircraft - X-1 Flight Summary," NASA, accessed February 2, 2016, http://www.nasa.gov/centers/dryden/history/HistoricAircraft/X-1/fltsummary.html.

[5] Robert Stanley, Bell Aircraft Corporation memorandum of May 11th, 1945.

[6] http://www.nasa.gov/centers/armstrong/news/FactSheets/FS-079-DFRC.html 18 February 2016. Stan Smith meeting notes from 1945.

[7] "Interview With Frank K. "Pete" Everest Who Flew A Bell X-2 To Record Speed of Mach 3," Aviation History, accessed February 18, 2016, http://www.historynet.com/aviation-history-interview-with-frank-k-pete-everest-who-flew-a-bell-x-2-to-record-speed-of-mach-3.htm.

[8] "The Bell X-2 'Starbuster,'" 456th Fighter Interceptor Squadron, accessed February 15, 2016, http://www.456fis.org/BELL_X-2_STARBUSTER.htm.

[9] "X-2," Encyclopedia Astronautica, accessed February 19, 2016, http://www.astronautix.com/craft/x2.htm.

[10] Robert W. Kempel and Richard E. Day, "A Shadow Over the Horizon, The Bell X-2," *American Aviation Historical Society Journal*, Spring 2003, accessed February 2016, http://www.bellx-2.com/sightings/horizon/article.html.

[11] A. Scott Crossfield with Clay Blair, Jr., *Always Another Dawn: The Story of a Rocket Test Pilot* (Cleveland: World Publishing Co., 1960)

[12] Bell Aircraft Corporation Internal Report 50-46

[13] Peter W. Merlin and Tony Moore, *X-Plane Crashes: Exploring Experimental, Rocket Plane, and Spycraft Incidents, Accidents and Crash Sites* (Forest Lake: Specialty Press, 2008)

[14] Frank K. Everest, *The Fastest Man Alive* (New York: Berkley Publishing Group, 1986)

[15] "Iven Carl Kincheloe, Jr.," Arlington National Cemetery Website, accessed February 24, 2016, http://www.arlingtoncemetery.net/kinchel.htm.

[16] "Interview With Frank K. "Pete" Everest Who Flew A Bell X-2 To Record Speed of Mach 3."

[17] "Kincheloe Memorial Park Rededication Ceremony", *Purdue Aerogram*, Fall 2007, page 19

[18] Kempel and Day, "A Shadow Over the Horizon."

[19] Peter W Merlin, "Starbuster: 55 Years Ago Capt. Mel Apt Conquered Mach 3, Lost Life on Fated Flight" NASA Dryden History Office, October 5th, 2011, accessed March 2nd, 2016, http://www.edwards.af.mil/News/Article-Display/Article/394758

[20] Kempel and Day, "A Shadow Over the Horizon."

[21] "NASA Armstrong Fact Sheet: Bell X-2 Starbuster," NASA, accessed March 2, 2016, https://www.nasa.gov/centers/armstrong/news/FactSheets/FS-079-DFRC.html.

[22] J. Terry White, "Final Flight: Bell X-2," *White Eagle Aerospace Blog*, September 28, 2015, http://www.whiteeagleaerospace.com/final-flight-bell-x-2/.

[23] "NASA Armstrong Fact Sheet: Bell X-2 Starbuster."

[24] "Iven C. Kincheloe Award Critera," SETP Website, accessed June 24, 2016, https://www.setp.org/criteria/iven-c-kincheloe-award.html.

AUTHOR INDEX